T0184554

Blockchain Platforms

A Look at the Underbelly of Distributed Platforms

Synthesis Lectures on Computer Science

The Synthesis Lectures on Computer Science publishes 75–150 page publications on general computer science topics that may appeal to researchers and practitioners in a variety of areas within computer science.

© Springer Nature Switzerland AG 2022

Reprint of original edition © Morgan & Claypool 2020

All rights reserved. No part of this publication may be reproduced, stored in a retrieval system, or transmitted in any form or by any means—electronic, mechanical, photocopy, recording, or any other except for brief quotations in printed reviews, without the prior permission of the publisher.

Blockchain Platforms: A Look at the Underbelly of Distributed Platforms
Stijn Van Hijfte

ISBN: 978-3-031-00676-0   paperback
ISBN: 978-3-031-01804-6   ebook
ISBN: 978-3-031-00072-0   hard

DOI: 10.1007/978-3-031-01804-6

A Publication in the Springer series
SYNTHESIS LECTURES ON COMPUTER SCIENCE
Lecture #11

Series ISSN 1932-1228 Print 1932-1686 Electronic

Blockchain Platforms

A Look at the Underbelly of Distributed Platforms

Stijn Van Hijfte, Howest Applied University College

SYNTHESIS LECTURES ON COMPUTER SCIENCE #11

ABSTRACT

This book introduces all the technical features that make up blockchain technology today. It starts with a thorough explanation of all technological concepts necessary to understand any discussions related to distributed ledgers and a short history of earlier implementations. It then discusses in detail how the Bitcoin network looks and what changes are coming in the near future, together with a range of altcoins that were created on the same base code. To get an even better idea, the book shortly explores how Bitcoin might be forked before going into detail on the Ethereum network and cryptocurrencies running on top of the network, smart contracts, and more. The book introduces the Hyperledger foundation and the tools offered to create private blockchain solutions. For those willing, it investigates directed acyclic graphs (DAGs) and several of its implementations, which could solve several of the problems other blockchain networks are still dealing with to this day. In Chapter 4, readers can find an overview of blockchain networks that can be used to build solutions of their own and the tools that can help them in the process.

KEYWORDS

Blockchain, distributed ledger, DAG, Bitcoin, Ethereum, Iota, Hyperledger

Contents

Introduction

Why another book on blockchain? I asked myself the same question when I started to write this very line. The reason is actually quite simple. By this point, everyone seems to have heard of blockchain in one way or another. But it is clear to me that, on average, not one person really understands the core concepts or really knows what it was all about. Others had deep knowledge but either only of the core concepts or of one specific platform. Theory and personal perception are the core of the information and many of the sources I scoured from all over the world all seemed to be limited. Not bad, just limited. They all brought only a small piece of the puzzle that I had to try to form myself.

This book attempts to give an extensive explanation of the core concepts and explain several platforms. Do I want to create a training manual that explains each platform in excruciating detail? No. However, this book should at least show you how these platforms generally work and, upon your choosing, allow you to investigate certain platforms in more detail on your own. All of the information is out there; it is up to you to go out and explore!

CHAPTER 1

Underlying Concepts and Technologies

To understand blockchain and distributed ledger solutions, one should also understand the underlying technologies and concepts that make up, or at least have influenced, the creation of blockchain technology. Only that way can we hope to understand how the idea of Bitcoin and blockchain saw the light of day. You can simply pass through this section, but you will see that if you are willing to take the time to understand what underlying principles have helped to define blockchain, it will give you an edge once we start to dive deeper into blockchain itself.

1.1 HASH FUNCTIONS

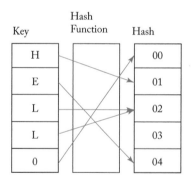

Figure 1.1: Hashing.

The use of hash functions or "hashing" is the use of a certain algorithm to map an input of variable-length arbitrary input data to a unique fixed-length output. This output is what we generally also call a "hash." Important to understand is that these are one-way functions, meaning that you shouldn't be able to reverse the output data back to the original input data. We can often make the link between these hashes with "hash tables," which are simply tables containing a number of hashes to speed up the look up of a certain hash in computer software. So here are some of the most important characteristics of cryptographic hash functions.

- **Pre-image resistance:** for a certain hash value a it should be difficult to find any other message m so that hash$(m) = a$.

- **Second pre-image resistance:** when we consider a message m_1, it should be difficult to find a second message m_2 so that hash(m_1) = hash(m_2).

- **Collision resistance:** it should be difficult to find to different values that lead to the same hash value.

Hash functions have some specific characteristics which make them so useful in today's world, both for cryptography and for blockchain platforms. One of these relations is that a specific input in the function will always lead to a specific output. If you only change one letter, number, or symbol, the entire output will change; this has three valuable properties. First of all, it allows us to check the integrity of what you have received. For example, if you download certain software on your PC, you can often check if the hash of your download matches the hash of the website. When this is not the case, you know there is a possibility that you have downloaded malware or at least not the application you were originally looking for. Second, when you receive the hash, you have no way of knowing what the input was. As we said before, only changing one letter will lead to a completely different outcome. In the world of cybersecurity there is something called "rainbow table attack." In this case the attacker has a table of hashes of which he knows the input and the specific hash function that has been used and he compares these with the hash of a password he was able to steal. If he finds a match, he knows your password. So if an attacker knows which hashing algorithm was used, he can try to guess what the input was but normally the hash itself will not help him with this (some hashing algorithms have been "broken" in the past so that attackers can much faster and easier guess what the original input was). The final positive outcome we are looking for is a function that is collision-resistant, leading to a situation where you cannot find two inputs that have the same output. Ultimately, every hashing function is not completely collision resistant because there is a variable input and only a limited fixed output. At least it should take a long time to find such a collision, otherwise the function loses its value. Over time, several hash functions have commonly been in use within software. The reason why these have been developing over time is because with increasing computer power, several of the "older" standards have been broken, while others contain vulnerabilities that have been exploited. Such attacks include the "collision attack" where the attacker will try to find two inputs which lead to the same output. No hash function is really "collision-free" but it should normally take an extremely long time to find such a collision. Other attacks are the "birthday attack" and the "pre-image attack." There are even competitions that aim at breaking current hash functions such as SHA-256. Why? When vulnerabilities are discovered, better hash functions are created that overcome these vulnerabilities. Wouldn't you rather have that an attacker breaks the function "legitimately" than using his newfound knowledge for malicious activities?

1.2   ECASH

ECash was an idea of David Chaum's back in 1983.[1] It was meant as an anonymous cryptographic way of spending money where the money would be stored on the computer of the user and signed by the software of the bank. There was a public key system but the bank was in control as a central authority. Eventually in 1990 DigiCash implemented the system with Mark Twain Bank in Saint Louis where over three years about 5,000 people made use of the system. It would remain in use until 1998 (even though DigiCash went broke in 1993). ECash did know some use in Europe at several institutions as electronic payments started to become more and more popular and organizations were looking for new ways of performing transactions. The reason why David Chaum deserves specific credit here is that his paper from 1982 "Computer Systems Established, Maintained, and Trusted by Mutually Suspicious Groups" had all elements that we would see in the first blockchain implementation, except for the proof of work algorithm.[2]

1.3   HASHCASH

The second of the earlier technologies that preceded Bitcoin and other blockchain platforms we would like to reference is hashcash which is a proof of a work algorithm which has been used as a counter measure for denial-of-service attacks in several systems. It was invented by Adam Back in 1997 and is to this day widely used all over the world as it is being used as the bitcoin mining function.[3] The cost function must be efficiently verifiable, but expensive to compute. In human terms this means it must be easy to verify if a found solution is also the correct solution, while it should be sufficiently hard to find such a solution in the first place.

1.3.1   THE HASHCASH COST FUNCTION

The hashcash cost function is used in the Bitcoin network to this day. It is a non-interactive, publicly auditable, trapdoor-free cost function with unbounded probabilistic cost. What does this actually mean? Publicly auditable means that the cost function can easily be verified without making use of any secret information or trapdoors.[4] There should be no trapdoors within the cost function we are using, otherwise the protocol itself could be broken with anyone who has knowledge about the function and one could no longer rely on the fact if one actually did the work or achieved the result by making use of this trapdoor. It would attack the trust that exists within the network and

---

[1]   Edwin (November 15, 2017). 1983: eCash door David Chaum. https://www.bitcoinsaltcoins.nl/1983-ecash-david-chaum/. Accessed May 17, 2020.

[2]   Wat is de geschiedenis van blockchain? https://btcdirect.eu/nl-nl/geschiedenis-blockchain. Accessed June 3, 2019.

[3]   The paper on Hashcash can be found here: http://www.hashcash.org/papers/hashcash.pdf.

[4]   We do not mean literally "auditable" by an auditor that would repeat the work in any sense. It means that we can efficiently verify the result of a cost function without the cost of repeating the work to come to this final result.

be detrimental to the willingness of the miners to invest time and energy to solve the challenge. As we noted earlier, the result of a cost function of any kind should be easy to verify but difficult to create in the first place.[5] But we can understand that there are different types of cost functions. A first difference can be made between fixed cost and probabilistic cost functions. The fixed cost function, as you could have guessed makes use of a fixed amount of resources to compute, with the fastest possible algorithm a deterministic one. A probabilistic cost function, on the other hand, has a predictable expected time to compute but in reality has a random actual time because the client is trying to compute the cost function by making use of a randomly selected start value.[6] Within the probabilistic cost functions we can again make a distinction between two groups: the unbounded and bounded probabilistic cost function. An unbounded cost function can in theory take forever to compute, although one should note that the probability of taking longer than expected quickly decreases toward zero. With a bounded cost function, one should know and realize that there is a limited space within the result can be calculated (rings a bell?), a specific key space needs to be searched so there is always an upper bound to the cost of finding a solution:

$$\begin{cases} \mathcal{C} \leftarrow \text{CHAL}(s, w) \text{ server challenge function} \\ \mathcal{T} \leftarrow \text{MINT}(\mathcal{C}) \text{ mint token based on challenge} \\ \mathcal{V} \leftarrow \text{VALUE}(\mathcal{T}) \text{ token evaluation function} \end{cases}$$

We see above a challenge c called by the server toward the client, computed by the server by making use of the CHAL() function where the service-name bitstring s and the amount of work w are the key parameters. The client then has to compute a token $\mathcal{T}$ using a cost function MINT()[7] and with the work difficulty w as part of the challenge. Finally, the server will check the token by making use of an evaluation VALUE(). There is also a possibility for a non-interactive cost function (so without the interaction between the server and the client) where the client can choose a challenge or a random start value in the MINT() function:

$$\begin{cases} \mathcal{T} \leftarrow \text{MINT}(s, w) \text{ mint token} \\ \mathcal{V} \leftarrow \text{VALUE}(\mathcal{T}) \text{ token evaluation function} \end{cases}$$

For the hashcash cost function, you should take some consideration for the following notation, introduced by Adam Back. Considering a bit string $s = \{0,1\}^*$, we can define $[s]_1$ so that this is the left-most bit while $[s]_{|s|}$ is the right-most bit so that $s = [s]_{1..|s|}$. There is also the binary infix comparison operator $x \overset{\text{left}}{=} b$ where b is the length of the common left-substring of the two bitstrings:

$$x \overset{\text{left}}{=}_0 y[x]_1 \neq [y]_1$$

[5]   Back, A. (August 1, 2002) *Hashcash – A denial of Service Counter-Measure.*

[6]   You can sense that "luck" is also an important factor here, as it is almost a lottery to find the right starting value within the cost function. This is also the case for the challenge put forward by the Bitcoin protocol.

[7]   The "mint" refers to the analogy between creating a cost token, or cryptocurrency, and actually minting physical money.

$$x \overset{\text{left}}{=}_b y \forall_{i=1\ldots b}[x]_i = [y]_i$$

We see that the hashcash cost function is computed relative to a service-name s to prevent tokens minted from another server being used on another. This service-name s can be any bit-string which is used to uniquely identify the service (such as the host name or an email address). The basis of the cost function is finding partial hash collisions on all 0 bits k-bit string 0^k and the fastest way to do this is by making use of brute force. To make sure there is no double spending the server (or network) needs to keep a ledger of all transactions so that it is clear that all actors remain true. There should also be a time constrain to take into account clock inaccuracy, computation time and transmission delays.

$$\begin{cases} \text{PUBLIC} & \text{hash function } \mathcal{H}(\cdot) \text{ with output size } k \text{ bits} \\ \mathcal{T} \leftarrow \text{MINT}(s, w) & \textbf{find } x \in_R \{0, 1\}^* \textbf{ st } \mathcal{H}\,(s\|x) \overset{\text{left}}{=}_w 0^k \\ & \textbf{return } (s, x) \\ \mathcal{V} \leftarrow \text{VALUE}(\mathcal{T}) & \mathcal{H}\,(s\|x) \overset{\text{left}}{=}_v 0^k \\ & \textbf{return } v \end{cases}$$

In practice, the value of $|x|$ could be chosen large enough (around 128 bits could suffice depending on the use case) to reduce the probability that a previously used value us reused by the client. The server can retain a double-spending database with a timestamp to discard entries from the spent database after they have expired. The interactive hashcash cost function is used in interactive settings where we see TCP, TLS, SSH, etc. In the original implementation it was used as a challenge chosen by the server to defend its resources against DoS-attacks.

$$\begin{cases} \mathcal{C} \leftarrow \text{CHAL}(s, w) & \text{choose } c \in_R \{0, 1\}^k \\ & \textbf{return } (s, w, c) \\ \mathcal{T} \leftarrow \text{MINT}(C) \, \textbf{find } x \in_R \{0, 1\}^* \textbf{ st } \mathcal{H}\,(s\|c\|x) \overset{\text{left}}{=}_w 0^k \\ & \textbf{return } (s, x) \\ \mathcal{V} \leftarrow \text{VALUE}(\mathcal{T}) & \mathcal{H}\,(s\|c\|x) \overset{\text{left}}{=}_v 0^k \\ & \textbf{return } v \end{cases}$$

Several improvements were proposed over the years, where, i.e., the target string to find the hash collision against a fixed output string as it is simpler and reduces verification cost. Even though it is used in Bitcoin, several important aspects have been changed: the hashing algorithm (SHA-1 vs. SHA-256, 20–160 hash bits 0 vs. at least first 32 of 256 hash bits zero, Bitcoin periodically resets difficulty level as explained later).

1.4   B-MONEY

B-money was created by Wei Dai in an effort to create an anonymous, distributed electronic cash system in a paper that was published on the cypherpunks mailing list in 1998. He proposed two separate protocols to come to his solution. The first was a symmetric proof of work function to create his digital currency. The reason why this protocol wouldn't be deemed acceptable was because that it required a broadcast channel that could not be jammed and remained synchronous. The B-money is transferred by broadcasting all transactions to all participants. All the participants are hereby forced to keep accounts on all other participants. When there are conflicts on the network, each party can broadcast the evidence over the network and each participant needs to determine the effect and outcome for themselves within the accounts they keep. As you can sense this is not something that would be sustainable over time both from a technical standpoint (broadcasting, un-jammable, synchronous) as from a business standpoint (determination by each participant of what the eventual outcome is of a dispute within accounts). A second approach was based on a small set of participants in the network that would keep the accounts ("servers"). They would have to lock a certain amount of money to become a server and lose it if they proved to be dishonest. Important was that the other participants had to keep on checking the accounts to make sure the servers remained honest and to verify that the money supply was not being affected by this small group of participants. The importance of B-money and the ideas behind it cannot be underestimated. Satoshi Nakamoto, the writer of the Bitcoin paper, even made a reference to B-money. [8]

1.5   PEER-TO-PEER

A peer-to-peer network is a different way of working when it comes to the "classic" internet. In the usual way of accessing the internet, when a person tries to access a web page, a request is sent from your computer to a central server where the web page resides. The server responds and you see the webpage on your client. Every person on the planet that tries to access the same webpage, will send a request to the same central web server (more or less, but we want to keep the example as simple as possible).

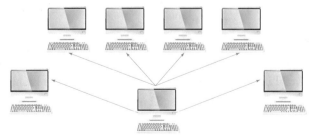

Figure 1.2: Server–client environment.

[8]   If you would like to know more, please do not hesitate to read the paper written by Wei Dai: http://www.weidai.com/bmoney.txt.

In a peer-to-peer (P2P) environment, the clients are connecting directly to each other and transfer data between one and another without a central server taking any part in the process. P2P is very similar to how humans communicate in real life: several nodes need to identify each other directly (making use of the IP-address of the partner) and check if communication over the correct port is possible. Only if this is the case, communication can begin. P2P networks are different from other distributed systems on the following aspects. [9]

- **Symmetric role:** each participant typically acts as a client and a server.

- **Scalability:** there is no all-in-all communication in the network necessary, allowing for scaling.

- **Heterogeneity:** node software can run on different hardware.

- **Distributed control:** control in the network is shared by all participants.

- **Dynamism:** P2P applications are often functional in a very dynamic environment.

There are centralized possibilities for P2P systems but these systems lack the robustness and scalability one can expect from an actual decentralized P2P system. Other specifications that one needs to take into account for decentralized systems are the following: the network can either be hierarchical or flat, and the overlaying network can either be structured or unstructured. The first concept relays to the fact that there can be a network where certain nodes have more responsibilities than other nodes in the network (often called "supernodes" or "masternodes"). We will see several blockchain networks where this is actually the case. This can deliver several advantages to a network, such as faster synchronization, better message throughput, and increased scalability. These special nodes can become responsible for voting on events in the network or carry other key responsibilities which help determine the future of the network. Second, there is the choice between structured and unstructured data. In case we are working with unstructured data, each of the peers is responsible for its dataset and the queries on how it can forward this information. This comes with several disadvantages because it makes it very difficult to return data from such a network, which will often be incomplete and you are never sure that you will find back the correct data. Structured networks, on the other hand, work with a predefined data format (specific JSON-format, XML-format, etc.). This allows for better information transferring and you can predict how the data will look like, once queried from the network. These types of networks come also with a set of challenges. The system can be very unpredictable as all nodes in the network control their own actions and can decide to enter or leave the network whenever they want. This, in turn, has an effect on the performance of the network. Depending on the nodes, the same query for data can be answered by multiple nodes

[9]   Vu, Q.H., Lupu, M., and Ooi, B.C. (2010). *Peer-to-Peer Computing: Principles and Applications.* 1st ed. Springer, p. 35.

or not at all, but also at different cost vectors. As data is copied over multiple nodes, there is the possibility that some of the nodes have conflicting data sets or even outdated data. There is no central point of a single truth, which can bring problems of its own. The networking aspect of a P2P system can also provide challenges as queries need (in some cases) to be broadcasted throughout the network, leading to congestion when there are a lot of these requests. Furthermore, the network cannot always support complex queries (as this means that nodes need to process more information on the data they are storing); security is often in question as the network is often open to all participants, privacy is a concern as there is accountability for the actions of a node in the network, and there is the need for incentives so that participants will actually follow the rules of the network and support it. Finally, there is the need of a parallel programming model (which such networks sadly often lack). Routing in such networks (as stated before) often poses a challenging task. Certainly, the scalability of unstructured P2P networks often poses a challenge because of network congestion. This is the trade-off one has to make for high autonomy and low maintenance costs of the nodes. Often, these networks make use of TTL (time-to-live) to determine the validity of queries that are being propagated throughout the network. Several techniques for routing have been defined over time. First, we are going to give a short overview of the techniques used in unstructured networks: breadth-first search (BFS) and depth-first search (DFS). The first has a predefined parameter D that determines the maximum TTL of a query. It forwards the query to all nodes and the message keeps on propagating until D is reached. The second technique, on the other hand, also uses the same parameter D but only sends the query to the most promising nodes in the network. There are also heuristic-based routing strategies such as iterative deepening. Iterative deepening is a technique that has been derived from AI research and basically comes down to several breadth-first searches that are been scaled over time until the TTL is reached (or the result is returned). Another technique is directed BFS, where the message is first propagated to a subset of the neighboring nodes, after which "classic" BFS takes over. A third technique is called "local indices search" where the node not only creates an index for its local data but also for the data stored at neighboring nodes. Routing indices-based search takes it up a notch by making nodes store topics and the number of documents stored by their neighbors. There are several techniques that fall under routing indices-based search (RI) called compound RI, hop-count RI, and exponential aggregated RI. Next, there is the random walk where the query is forwarded to one or more random neighbors until the result is found or TTL is reached. The sixth technique is adaptive probabilistic search which is a random walk query with some probabilities added. The node sending the message will choose its neighbor based on some indicators. Next, is the bloom filter-based search which will make use of bloom filters (explained later) to determine which information is stored at which (neighboring) node. Structured P2P networks also have a number of routing options. These can be divided in three groups: networks using distributed hash tables, skip list systems, and tree-based systems. The first uses distributed hash tables (explained below), the second will make use of a skip graph or something similar which are lists and the nodes participate at each level of the lists. The tree-based systems make use of a tree-based structure to index the data. The first routing technique used is called "chord" and makes use of hashes to map nodes and data items in a single-dimensional identifier space. Another is "CAN" or "content addressable network" which is built on a virtual d-dimensional Cartesian coordinate space. A third type is called "PRR trees" of which "Pastry" and "Tapestry" are two implementations. Each node constructs a routing table based on node-IDs and these IDs indicate which of them are closest. A third technique is called "Viceroy" which is a multi-level DHT-based system while a fourth called "Crescendo" takes a hierarchical DHT approach. Skip graph, as described above, makes use of multiple lists at a level. There is also SkipNet, P-Grid, P-Tree, and BATON. There are of course also hybrid P2P systems which make use of techniques such as "ultrapeers." In this technique there are two types of nodes: ultrapeers and leaf peers. The first will forward queries to other ultrapeers while these ultrapeers will also search its leaf peers based on indexing to locate the node with the desired data. There is also the system of structured supernodes and edutella. The concept of "trust" is also an important one to consider when we talk about P2P systems. How do you trust the other nodes in the network? In a world without servers there are two techniques that can be used to manage trust: a gossip protocol to exchange knowledge among the all the nodes in the network while the second techniques only focuses on "local" reputation. One of the first P2P networks was Napster (a centralized P2P network), which allowed file sharing over the internet. If we are really looking at fully decentralized P2P networks, we know examples such as FreeNet, Gnutella, FreeHaven, and others. The BitTorrent protocol is another such a specific implementation of a P2P network that is still in use today.

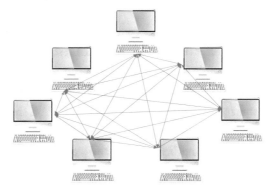

Figure 1.3: Peer-to-peer.

In the BitTorrent world, a computer joins a "swarm" when it starts loading a .torrent file. Based on this file, the client contacts a tracker, which is a server that keeps track of all the connected clients. The clients start to share IP addresses based on this information, after which data can be shared. From the second you have downloaded a specific bit of a file, you start to upload it as well in the network, speeding up every participant's download speed without stressing a single machine. This tracker system is being replaced more and more by a traceless system so that a central server can be completely avoided. This is done by making use of a distributed hash table. From the second you start using a "magnet link" more and more nodes in the network are contacted until the information on the file requested is found.

1.6   DISTRIBUTED HASH TABLE

A distributed hash table (DHT) is a hash table in a decentralized distributed system. This means that key-value pairs are stored in a table and each node in the network can make use of the key to look up a certain value in the table. The nodes in the network are responsible to keep up the table with unique keys so that there are no double values and hence issues in the network. These DHTs came into existance by P2P networks such as Napster, Freenet, Gnutella, and BitTorrent. These DHTs are easily scalable with any number of nodes in the network. They are also fault tolerant and easily maintained as it doesn't matter as new nodes join or leave the network. When we go into more detail, we already know that the DHT consists out of a key. This keyspace partitioning scheme is divided among all the participating nodes. There is also an overlay network that is responsible for connecting all the nodes (structured). The final component is the actual hashing of the values in the table. There is either consistent hashing or rendezvous hashing that helps to map keys to nodes. The advantage here is that only neighboring nodes are influenced when a node enters or leaves the network.

1.7   DECENTRALIZED VS. DISTRIBUTED

We often talk about a distributed network vs. a decentralized network but what is the difference between these two and how do they differ from classic centralized solutions? One can easily imagine a centralized structure. As in the example with the server and the clients in the paragraph before, you have one point of failure and one point of control. This is the central power in the network that determines everything. Decentralized is the next step. Here there isn't a single central power, but you can find several of these "central" clusters within the network. So, the network has moved away from a central, single point of failure but we haven't reached the stage of a "distributed" network just yet. In a distributed network no central point(s) can be discovered in the network, so taking out a node will not lead to information loss as there is no single point of failure (nor, i.e., four points of failure).

This structure can be seen as a subset of the distributed network where power between all the nodes is even more divided throughout the network. There is no longer a real point of failure as any node can leave or enter the network without impacting the overall functionality (up to a certain point depending on the replication rate of date throughout the network).

Figure 1.4: Decentralized structure.

1.7.1   SHA-256

Secure hashing algorithms (SHA) are a specific family of cryptographic functions which were specifically designed to deal with the security of data. And as you guessed it, they do so by making use of hashing. What is hashing? Easily explained it is a mathematical algorithm to come to a fixed-size output.[10] No matter how long the input is, the output will always have the same size! Change one letter, one number, or even something as small as a dot (.) and the output will be completely transformed. Or at least that is how a good functioning hashing algorithm works.[11] It will make use of Boolean operations such as AND ($\wedge$), XOR ($\oplus$), and OR ($\vee$), bitwise complements (¯) and integer addition modulo 2^{32} (at least for SHA-256). Several attacks are imaginable on these hashing algorithms. The first is a brute force attack where the attacker tries every possible input to find the output. The harder it is for the attacker to find the original message, the better the so-called "pre-image resistance." A second type of attack is the collision attack. Here the attacker tries to find two possible inputs that lead to the same kind of hashing output. If this happens, and an application compares the hashes rather than the passwords, an attacker could gain access without actually knowing the correct password. A third type of attack is the birthday attack which is based on the "birthday paradox." This paradox states that, if you are in a given room with for example 30 people, and you ask everyone for their birthday, you have about 70% chance that 2 of them will share the same birthday. Not to go too deeply into any explanation,[12] the birthday attack the attacker will try to create the same hash as the correct one but with incorrect input. This can be done when the output string isn't large enough. Then there are only a limited amount of possibilities of how input is transformed to output. Of course, you could say that this is also impossible, because the output of a hashing function is always limited. This is true, but the larger the output becomes, the more difficult it will be for the attacker to perform this kind of attack. A final attack that could take place

---

[10]   Important to note is that the output of a hashing algorithm is also one-way, which means it is impossible to transform the output back to the original input data. Or at least that is the theory…

[11]   There have been hashing algorithms in the past that have been broken and can now be reverse engineered.

[12]   The formula behind the paradox is the following: $1 - 365!((365 - n)! * 365^n)$ with n the number of participants in a room.

is the rainbow table attack. A rainbow table is a table that simply contains a list of inputs with their hashed outputs. All the attacker does is trying to match a certain hash to a specific input. There are several types and families of hashing algorithms but the focus here lies on the SHA-family. Over time there has been SHA-0, SHA-1, SHA-2, and SHA-3. SHA-0 and SHA-1 were proven to be vulnerable to several types of attacks in the past. SHA-2 produces 224 or 256 sized digests where SHA-1 only produced 160-bit digests. SHA-256 has a digest length of 256 bits and is a keyless hash function. Its algorithm operates on 32-bit words and makes use of the following functions (just to give you a general idea):

$$Ch(X, Y, Z) = (X \wedge Y) \oplus (\overline{X} \wedge Z),$$
$$\mathrm{Maj}(X, Y, Z) = (X \wedge Y) \oplus (X \wedge Z) \oplus (Y \wedge Z),$$
$$\Sigma_0(X) = RotR(X, 2) \oplus RotR(X, 13) \oplus RotR(X, 22),$$
$$\Sigma_1(X) = RotR(X, 6) \oplus RotR(X, 11) \oplus RotR(X, 25),$$
$$\sigma_0(X) = RotR(X, 7) \oplus RotR(X, 18) \oplus ShR(X, 3),$$
$$\sigma_1(X) = RotR(X, 17) \oplus RotR(X, 19) \oplus ShR(X, 10),$$

where $RotR(A, n)$ stands for the circular right shift, $ShR(A, n)$ stands for the right shift of the binary word A, and $A\|B$ stands for the concatenation of both the binary words A and B. It also makes us of the 64 binary words K_i given by the first 32 bits of the fractional parts of the cube roots of the first 64 prime numbers:

0x428a2f98	0x71374491	0xb5c0fbcf	0xe9b5dba5	0x3956c25b	0x59f111f1	0x923f82a4	0xab1c5ed5
0xd807aa98	0x12835b01	0x243185be	0x550c7dc3	0x72be5d74	0x80deb1fe	0x9bdc06a7	0xc19bf174
0xe49b69c1	0xefbe4786	0x0fc19dc6	0x240ca1cc	0x2de92c6f	0x4a7484aa	0x5cb0a9dc	0x76f988da
0x983e5152	0xa831c66d	0xb00327c8	0xbf597fc7	0xc6e00bf3	0xd5a79147	0x06ca6351	0x14292967
0x27b70a85	0x2e1b2138	0x4d2c6dfc	0x53380d13	0x650a7354	0x766a0abb	0x81c2c92e	0x92722c85
0xa2bfe8a1	0xa81a664b	0xc24b8b70	0xc76c51a3	0xd192e819	0xd6990624	0xf40e3585	0x106aa070
0x19a4c116	0x1e376c08	0x2748774c	0x34b0bcb5	0x391c0cb3	0x4ed8aa4a	0x5b9cca4f	0x682e6ff3
0x748f82ee	0x78a5636f	0x84c87814	0x8cc70208	0x90befffa	0xa4506ceb	0xbef9a3f7	0xc67178f2

The algorithm also makes use of padding to make sure that the input has a length that is an exact multiple of 512 bits. It always follows a specific procedure whereby a bit 1 is appended, k bits 0 are appended,[13] and the length $l < 2^{64}$ of the input is represented by exactly 64 bits. These are all added to the end of the message. For each block $M \in \{0,1\}^{512}$, 64 words of 32 bits are constructed with the first 16 being created by splitting M in 32-bit blocks $M = W_1\|W_2\|\ldots\|W_{16}$ and the remaining 48 are created by: $W_i = \sigma1(W_i{-}2) + W_i{-}7 + \sigma0(W_i{-}15){+} W_i{-}16$, $17 \leq i \leq 64$. The initial step of the hash computation is done by setting 8 variables to their initial values, which are given by the first 32 bits of the fractional part of the square root of the first 8 prime numbers:

$$H_1^{(0)} = \mathrm{0x6a09e667} \quad H_2^{(0)} = \mathrm{0xbb67ae85} \quad H_3^{(0)} = \mathrm{0x3c6ef372} \quad H_4^{(0)} = \mathrm{0xa54ff53a}$$
$$H_5^{(0)} = \mathrm{0x510e527f} \quad H_6^{(0)} = \mathrm{0x9b05688c} \quad H_7^{(0)} = \mathrm{0x1f83d9ab} \quad H_8^{(0)} = \mathrm{0x5be0cd19}$$

[13] With k the smallest positive integer so that $l + 1 + k = 448 \bmod 512$ with l = length in bits of the input.

The initial formula is set to the following:

$$(a, b, c, d, e, f, g, h) = (H_1^{(t-1)}, H_2^{(t-1)}, H_3^{(t-1)}, H_4^{(t-1)}, H_5^{(t-1)}, H_6^{(t-1)}, H_7^{(t-1)}, H_8^{(t-1)})$$

What follows is the processing of each of the 64 blocks MI and do the following 64 times:

$$T_1 = h + \Sigma_1(e) + Ch(e, f, g) + K_i + W_i$$
$$T_2 = \Sigma_0(a)\ h = gaj(a, b, c)$$
$$h = g$$
$$g = f$$
$$f = e$$
$$e = d + T_1$$
$$d = c$$
$$c = b$$
$$b = a$$
$$a = T_1 + T_2$$

Out of which we can compute the new hashed values by:

$$H_1^{(t)} = H_1^{(t-1)} + a$$
$$H_2^{(t)} = H_2^{(t-1)} + b$$
$$H_3^{(t)} = H_3^{(t-1)} + c$$
$$H_4^{(t)} = H_4^{(t-1)} + d$$
$$H_5^{(t)} = H_5^{(t-1)} + e$$
$$H_6^{(t)} = H_6^{(t-1)} + f$$
$$H_7^{(t)} = H_7^{(t-1)} + g$$
$$H_8^{(t)} = H_8^{(t-1)} + h$$

The final step is the following:

$$H = H_1^{(t)} \| H_2^{(t)} \| H_3^{(t)} \| H_4^{(t)} \| H_5^{(t)} \| H_6^{(t)} \| H_7 \| H_{18}^{(t)}$$

and we end up with the results from the SHA256-algorithm. To give you an idea: *abc* translates to ba7816bf8f01cfea414140de5dae2223b00361a396177a9cb410ff61f20015ad. To give you an idea of how it works, I also added the entire process in Python:[14]

```
W  = 32       #Number of bits in word
M  = 1 << W
FF = M - 1 #0xFFFFFFFF (for performing addition mod 2**32)
```

[14] Smith, N.T. SHA 256 pseuedocode? *Stackoverflow.* https://stackoverflow.com/questions/11937192/sha-256-pseuedocode/46916317#46916317. Accessed May 26, 2020.

```
K = (0x428a2f98, 0x71374491, 0xb5c0fbcf, 0xe9b5dba5,
     0x3956c25b, 0x59f111f1, 0x923f82a4, 0xab1c5ed5,
     0xd807aa98, 0x12835b01, 0x243185be, 0x550c7dc3,
     0x72be5d74, 0x80deb1fe, 0x9bdc06a7, 0xc19bf174,
     0xe49b69c1, 0xefbe4786, 0x0fc19dc6, 0x240ca1cc,
     0x2de92c6f, 0x4a7484aa, 0x5cb0a9dc, 0x76f988da,
     0x983e5152, 0xa831c66d, 0xb00327c8, 0xbf597fc7,
     0xc6e00bf3, 0xd5a79147, 0x06ca6351, 0x14292967,
     0x27b70a85, 0x2e1b2138, 0x4d2c6dfc, 0x53380d13,
     0x650a7354, 0x766a0abb, 0x81c2c92e, 0x92722c85,
     0xa2bfe8a1, 0xa81a664b, 0xc24b8b70, 0xc76c51a3,
     0xd192e819, 0xd6990624, 0xf40e3585, 0x106aa070,
     0x19a4c116, 0x1e376c08, 0x2748774c, 0x34b0bcb5,
     0x391c0cb3, 0x4ed8aa4a, 0x5b9cca4f, 0x682e6ff3,
     0x748f82ee, 0x78a5636f, 0x84c87814, 0x8cc70208,
     0x90befffa, 0xa4506ceb, 0xbef9a3f7, 0xc67178f2)

#Initial values for compression function
I = (0x6a09e667, 0xbb67ae85, 0x3c6ef372, 0xa54ff53a,
     0x510e527f, 0x9b05688c, 0x1f83d9ab, 0x5be0cd19)

def RR(x, b):
    '''
    32-bit bitwise rotate right
    '''
    return ((x >> b) | (x << (W - b))) and FF

def Pad(W):
    '''
    Pad and convert
    '''
    mdi = len(W)% 64
    L = (len(W) << 3).to_bytes(8, 'big') #Binary of len(W)
    in bits
    npad = 55 - mdi if mdi < 56 else 119 - mdi #Pad so 64 | len;
    add 1 block if needed
```

```
    return bytes(W, 'ascii') + b'\x80' + (b'\x00' * npad) + L #64
    | 1 + npad + 8 + len(W)


def Sha256CF(Wt, Kt, A, B, C, D, E, F, G, H):
    '''
    SHA256 Compression Function
    '''
    Ch = (E and F) ^ (~E and G)
    Ma = (A and B) ^ (A and C) ^ (B and C) #Major
    S0 = RR(A, 2) ^ RR(A, 13) ^ RR(A, 22) #Sigma_0
    S1 = RR(E, 6) ^ RR(E, 11) ^ RR(E, 25) #Sigma_1
    T1 = H + S1 + Ch + Wt + Kt
    return (T1 + S0 + Ma) and FF, A, B, C, (D + T1) and FF, E, F, G


def Sha256(M):
    '''
    Performs SHA256 on an input string
    M: The string to process
    return: A 32 byte array of the binary digest
    '''
    M = Pad(M) #Pad message so that length is divisible by 64
    DG = list(I) #Digest as 8 32-bit words (A-H)
    for j in range(0, len(M), 64): #Iterate over message in chunks
    of 64
      S = M[j:j + 64] #Current chunk
      W = [0] * 64
      W[0:16] = [int.from_bytes(S[i:i + 4], 'big') for i in range(0,
      64, 4)]
      for i in range(16, 64):
        s0 = RR(W[i - 15], 7) ^ RR(W[i - 15], 18) ^ (W[i - 15]
      >> 3)
        s1 = RR(W[i - 2], 17) ^ RR(W[i - 2], 19) ^ (W[i - 2] >>
      10)
        W[i] = (W[i - 16] + s0 + W[i-7] + s1) and FF
        A, B, C, D, E, F, G, H = DG #State of the compression func-
      tion
```

```
    for i in range(64):
     A, B, C, D, E, F, G, H = Sha256CF(W[i], K[i], A, B, C,
     D, E, F, G, H)
    DG = [(X + Y) and FF for X, Y in zip(DG, (A, B, C, D, E, F,
    G, H))]
    return b''.join(Di.to_bytes(4, 'big') for Di in DG) #Convert
    to byte array

if __name__ == "__main__":
    bd = Sha256('Hello World')
    print(''.join('{:02x}'.format(i) for i in bd))
```

1.8   MERKLE TREE

A Merkle[15] tree or binary hash tree is a way of summarizing a large set of data on a sufficient manner because it will form a generalization of a hash chain. You could imagine an upside-down tree. We are dealing with a branching data structure, where every leaf node is labeled with the hash of a data block and every non-leaf node is labeled with the hash of the labels of its child nodes. The root of the tree can be found on the top, while the leave nodes will spread out underneath the structure. When we are dealing with a number of n data elements, you can check if a certain data element is part of the root with at most $2+\log_z(n)$ calculations, which proves that it is a very effective way of reducing data.

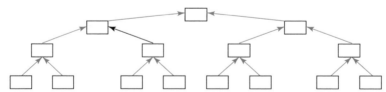

Figure 1.5: Simple example of a Merkle tree structure.

This technology can be found in P2P networks where it can be used to verify if data is being received undamaged. We can also find this within both the Bitcoin and the Ethereum networks. The Merkle trees help to summarize information in a sufficient manner; in the case of Bitcoin and Ethereum, it is the transactions that are being summarized. Hashes of transactions are being hashed together until there is only one hash left, the root, which will be included in the block header. When we are dealing with Bitcoin, the hashing algorithm used is double-SHA256. To verify if a transaction is included in a specific block, all that is needed for the node to identify this is the block

---

[15] The hash trees were created by Ralph Merkle who patented it in 1979.

header from which it can extract the Merkle root and consequently retrieve a Merkle path from the full node, without actually storing the entire blockchain. This is something we can find back in simplified payment verification systems (SPVs).

1.8.1 MERKLE PATRICIA TREE

When we talk about the Ethereum network, we make use of the Patricia Merkle Tree. This is a combination of the Merkle tree (described above) and the Patricia Trie. A Patricia Trie,[16] also called a Prefix Tree, Redix tree, or simple trie, is a data structure that uses a key as a path so that nodes that share the same key can share the same path. This means that this structure is the fastest for finding common prefixes and requires small memory at the same time.[17] When we make the step to the Merkle Patricia Tree, we see that each node receives a hash value which is decided by the SHA3 hash value of its contents. At the same time, this hash is used as the key that refers to this node. For example, Go-Ethereum uses levelDB while parity uses rocksDB to store states as key-value storages.[18] These key values of the Ethereum state are used as paths on the Merkle Patricia Tree. We express the unit to distinguish between the key values in "nibble." From this we know that each node can have up to 16 branches. When a node does not have a child, we call this a "leaf node." This node consists of two items: its path and its value. Next to the branch nodes and the leaf nodes we can also distinguish a third type of node: the extension node. This is an optimized node of the branch node. Within the Ethereum state, there are often branch nodes with one leaf node, which are than compressed to extension nodes which contain the path and the hash of the child. To be able to distinguish between leaf nodes and extension nodes, there is a prefix being added. If we are dealing with a leaf node which has an even number of nibbles, the prefix 0x20 is added; in the event of an uneven number of nibbles, the prefix 0x30 is added. For extension nodes, in the case of an even number of nibbles, we add 0x00 while in the case of an uneven number of nibbles we add 0x10.

1.8.2 BLOOM FILTERS

A final concept within the world of Merkle trees is the concept of Bloom filters. It is a probabilistic data structure that can tell with certainty if an item is not in the dataset while it can otherwise only state with a certain probability that a data item "might" be present in the dataset. This creates a situation where false negatives aren't possible while false positives can occur. What does this structure look like? It basically consists of a bit field and a set of hashing functions that eventually return a number with an index that corresponds with a bit in the bit field. So, if the Bloom filter is tested with an input, it can be sure if it has seen the input before when the bit within the bit field is 1,

---

[16]   Check out Medium.com for more information regarding the Patricia Merkle Trie or blockchain in general.
[17]   Commonly used for routing tables.
[18]   Keys and values saved in the storage are not the key-values of the Ethereum state.

otherwise it will be 0. How? The starting position of the bit field, is with all the bits turned to 0. When it is introduced to new data, certain bits are set to 1.

Table 1.1: Bloom filter set-up		
Starting Position	Introduce Elements X and Y	New Position
Bit Field		Bit Field
0		1
0	X	0
0		0
0	Y	0
0		1
0		1
0		0
0		0
0		0
0		1
0		1
0		0

Next, if you send data to the filter, it will know for sure that it has never seen the data before if not all the bits are equal to 1.

Table 1.2: Z is not in the set	
Is Z in the set	Bloom Filter
	1
Z	0
	0
	0
	1
	1
	0
	0
	0
	1
	1
	0

This is an interesting feature as Bloom filters are way more advantageous when it comes to space efficiency than other data structures because the data itself is not actually stored, only the results of the hashing of a data item is stored. In the world of blockchain this is sometimes used in wallet technology to speed up convergence (light clients).

1.9   STATE MACHINE

The concept of the "state machine" is an important one within computer science. It is a mathematical abstraction that is used to design algorithms.[19] The state machine reads in a series of inputs and with each input it will switch to a different state. Depending on the input, the state machine can tell you about the sequence of the input. Even though this might not sound very interesting at first, several problems within computer science can be solved with this concept. An example is the rendering of web pages. Even if you are not a programmer, you can imagine that a webpage needs to be rendered in a certain order. Otherwise there will be an error and you will receive nothing. The state machine can move the state over the several tags of a web page html file and make sure that everything is at least in the right order. A deterministic state machine is a state machine where that for each input there is only one transition, with the output being its final state. There is also the possibility of non-deterministic state machines. This means that several inputs come to the same transition within the state machine. This is possible because only the output is important when it comes down to a state machine. It is only at this moment that an external action is triggered. Below you can find an example of a finite state machine regarding you entering or exiting a car. There are only a certain set of possibilities that can lead to a certain set of states.

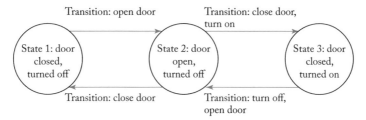

Figure 1.6: Finite state machine.

1.9.1   TURING MACHINE

A Turing machine is a hypothetical machine, namely one that is capable of recognizing non-regular patterns. It is computationally complete and anything that can be computed, can be computed on a Turing machine. It is not limited to a finite number of states, as it is able to make changes to the input it receives. We should also understand that it operates on an infinite amount of memory. The machine has a finite table with user-specified instructions which leave the machine to enter a

[19]   Shead, M. (February 14, 2011) State Machines – Basics of Computer Science. *Blog.markshead.com* https://blog.markshead.com/869/state-machines-computer-science/. Accessed June 5, 2019.

symbol and move on to the next input and instruction until it halts the computation.[20] A system of data-manipulation rules is called Turing complete, if it is capable of simulating a Turing machine. The system is able to recognize other data-manipulation rule sets. This way it shows the power of such a data-manipulation rule set.[21]

1.10   ELLIPTICAL CURVE CRYPTOGRAPHY

While the previous concepts and technological implementations might have been quite straight-forward to understand, this is certainly not the case for elliptical curve cryptography. I will try to explain the best way I possibly can, considering the many pitfalls one can face when trying to do so. First, I will have to explain in short what public key cryptography is. Public key cryptography (or asymmetric cryptography) is generally used to produce two types of keys: a private key and a public key. The public key can in some sense be shared with the public while the private key has to remain private (shocking, I know). To generate these keys a one-way function is used and came into practice to solve an age-old problem when it comes to secure communication. We all know symmetric key cryptography: you just produce a private key and use this as a password or passphrase for an account, lock, or anything else. Easy right? Imagine we want to send messages to one an-other, which we would like to encrypt so that we are the only people that can read it. Solution: we just share the same password, problem solved! Well, not really, because first of all we need to find a secure manner to share those keys with each other. This can already pose quite a problem. The second problem is the one of numbers. You might want to send secure messages to me, but I can imagine you would like to do the same with all your friends, family, and colleagues. That are a lot of private keys to send and store! And you haven't even started changing passwords or anything else. This would mean complete mayhem in the real world. Public key cryptography found the solution here. You can just share your public key with everyone, leave it in a public database and transmit it over insecure networks. It is meant to be shared and known by everyone. If you now use my public key to encrypt a message, I am the only person that can decrypt it with my private key.

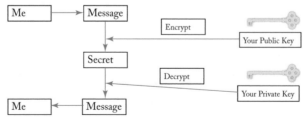

Figure 1.7: Public key cryptography.

[20]   Mullins, R. (2012) What is a Turing machine? Department of Computer Science and Technology – University of Cambridge. https://www.cl.cam.ac.uk/projects/raspberrypi/tutorials/turing-machine/one.html. Accessed June 5, 2019.
[21]   Invented by the well-known mathematician Alan Turing.

Some of you might have already made the link to blockchain and cryptocurrencies: your wallet address and the private key you use to access it. Of course there are many more uses for public key cryptography; think about digital signatures that are used to verify the sender of a message. The way these keys are generated is more often than not based on hashing algorithms that produce a certain outcome based on entropy. There is a reason why you can no longer access your wallet when you lose your private key: these are meant to be unbreakable. Of course when the algorithm can be broken, your keys can be broken as well. Depending on the blockchain platform you are using, different types of hashing are being used to generate these keys. This is also the reason why some coins can be stored together and others cannot. Often there is also some added procedures: this is why, i.e., the addresses for Ethereum and Bitcoin look so different, but more about that in the related sections. Now we can return to our original topic: elliptic key cryptography. This will bring us a bit out of our comfort zone (certainly if cryptography is completely new for you). To start with we have to talk about finite fields which can be defined as a finite set of numbers and two operations (addition and multiplication) that satisfy a specific set of rules.[22] These are:

- **Closed:** if a and b are in the set, so are $a + b$ and $a \cdot$ b;

- **Additive identity:** $a + 0 =$ a;

- **Multiplicative identity:** $a \cdot 1 = a$;

- **Additive inverse:** if a is in the set, so is $-a$ with $a + (-a) = 0$; and

- **Multiplicative inverse:** if a is in the set and is not $= 0$, so is a -1 with $a \cdot a$-1 $= 1$.

We can define a finite field the following way: $F_p = \{0, 1, 2, 3, \dots p\text{-}1\}$ where p stands for the order of the finite field F. The order will always be the power of a prime and this is because all the rules we have defined can only apply if the order is in fact a prime. If we want to remain consistent with the rules we have previously defined, we know that, i.e., addition will take on another form than regular addition if we wish to maintain the "closed" parameter. In finite fields, addition takes on the form of modulo arithmetic:

$$a +_f b \in F_p => a +_f b = (a + b)\% \, p \text{ with } a, b \in F_p$$

Similarly, we know that (where f denotes finite field operations such as subtraction, addition or multiplication):

$$a -_f b \in F_p => a -_f b = (a - b)\% \, p \text{ with } a, b \in F_p$$
$$-_f a = (-a)\% \, p$$
$$a \cdot_f b = a +_f a +_f a \ (b \text{ times})$$
$$a^b = a \cdot_f a \cdot_f a \ (b \text{ times})$$
$$n^{(p-1)}\% \, p = 1 \ (\text{Fermat's Little Theorem})$$

22   Song, J. (2019). *Programming Bitcoin: Learn How to Program Bitcoin from Scratch.* 1st ed. Boston, MA: O'Reilly, p. 123.

$$a / b = a \cdot_f (1 / b) = a \cdot_f b^{-1}$$

The next step are elliptical curves. These are of the form: $y^2 = x^3 + ax + b$,

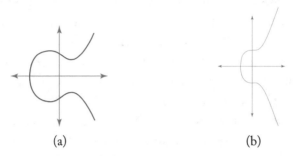

(a)   (b)

Figure 1.8: (a) Secp256k1 and (b) continuous elliptic curve.

These elliptical curves are used in cryptography and in blockchain implementations. For Bitcoin it is $y^2 = x^3 + 7$ or secp256k1. It is often said that this specific implementation was picked because it has the lowest probability of kleptographic backdoors being implanted by the NSA. This is why many other blockchain platforms make use of the same elliptical curve. But now toward the why: why do we need elliptical curves? Elliptical curves are used for something very specific: point addition. Point addition is actually just as it sounds. We add two points that lie on the curve. The weird thing is that the outcome of this addition, a 3rd point, will also be on the curve![23] This is a very interesting property that is thankfully being put to use. Also here there are severable properties that need to be respected:

- **Identity:** if $I = 0 \Rightarrow I + A = A$;

- **Commutativity:** $A + B = B + A$;

- **Associativity:** $(A + B) + C = A + (B + C)$; and

- **Invertibility:** $A + (-A) = I$.

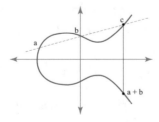

Figure 1.9: Point addition.

<hr />

[23]   Exceptions are when the intersecting line with the curve is perfectly vertical or when it is the tangent of the curve.

There is the case where $X_1 \neq X_2$. Here we have to base ourselves on the slope s of the curve to calculate (X_3, Y_3) where Y_3 is the reflection over the x-axis.

$$s = (Y_2 - Y_1) / (X_2 - X_1)$$
$$X_3 = s^2 - X_1 - X_2$$
$$Y_3 = s(X_1 - X_2) - Y_1$$

When $X_1 = X_2$ but $Y_1 \neq Y_2$ we have the situation where $P_1 + P_2 = I$

The reason behind this is because when $P_1 = P_2$ we will be dealing with the slope of the tangent when we start to do calculations and point addition will fail. Now we have come to the end of our journey: we will combine finite fields with elliptical curves. Most of the assumptions that we have defined for both finite fields and elliptical curves seem to work together without much of a problem. Only point division isn't that easy and is also called "the discrete log problem." It is this "problem" that is the foundation of elliptical curve cryptography:

$$P^a = Q \Rightarrow \log_p Q = a$$

The problem is that "logpQ" is no analytically calculable algorithm. Another aspect of elliptical curves over finite fields is that it gives us scalar multiplication which has an asymmetric problem as it is easy to calculate in one direction but hard to reverse. All of this is finally combined to come to the actual core of the business: elliptical curve cryptography for which we need finite cyclic groups. So what are groups? Groups are finite fields with only one operation, which in this case is point addition. We have to respect the following properties, with G being the generator point that helps us generate the group:

- **Identity:** $0 + A = A$;

- **Closure:** $(a + b)G = ((a + b)\% \, n)G$ with n the order;

- **Invertibility:** if aG is in the group, so is $(n - a)G$;

- **Commutativity:** $aG + bG = bG + aG$; and

- **Associativity:** $aG + (bG + cG) = (aG + bG) + cG$.

Now, if we wish to define the elliptical curve for public key cryptography, we will need to define the following information.

- What are a and b in $y3 = x3 + ax + b$?

- What is the prime p of the finite field?

- What is the generator point G?

- What is the order n of the group?

If you have all of this information, you can start creating your cryptographic curve which can be used in elliptical curve public key cryptography. It has many uses, but in the world of blockchain it is used for signing and verifying transactions. You can clearly see that it is crucial that these underlying curves cannot be broken by attackers, otherwise the security of the entire system would be at stake. Again, as mentioned before, secp256k1 was chosen by many blockchain platforms because it had the least chance of a backdoor being built in by the NSA. The parameters of this are the following:

- $a = 0, b = 7$, making the equation $y2 = x3 + 7$;

- $p = 2256 - 232 - 977$;

- Gx = 0x79be667ef9dcbbac55a06295ce870b07029bfcdb2dce28d959f2815b16f81798;

- Gy = 0x483ada7726a3c4655da4fbfc0e1108a8fd17b448a68554199c47d08ffb10d4b8; and

- n = 0xffffffffffffffffffffffffffffffffffffffffebaaedce6af48a03bbfd25e8cd0364141.

It is here that once again the discrete log problem shows himself. The equation that we need to solve is the following:

$$P = eG$$

Not that hard, is it? When we know e and G, it is easy as hell to compute P. However, if we know P and G, we cannot compute e! Welcome to the world of blockchain, where P is also better known as your public key and e is known as your private key.

1.11   BYZANTINE GENERALS PROBLEM

The Byzantine Generals Problem[24] is a famous problem in computer science that was solved by Satoshi Nakamoto. Important to note is that he was not the first person to solve the problem at hand, rather he gave another solution which proved to be successful. The original allegory is a thought experiment which tries to clearly explain the issues and challenges when you try to communicate in a distributed manner over unsecure or unreliable links. It refers to the Byzantine Empire where several armies try to coordinate an attack. The attack must happen simultaneously or the attack will fail and the armies will be defeated. The generals of these armies try to communicate with each other by sending messengers to one another. The problem is of course that a messenger can be killed, captured or bought. Even a general can turn on his comrades. They must both decide the time of attack and agree on this time. How will they reach consensus and succeed? And more importantly:

---

[24] Also known under other names, such as the "interactive consistency problem," which was coined by Robert Shostak. The Two Generals Problem was published in *Some constraints and Trade-offs in the Design of Network Communications* (1975) by Akkoyunlu, Ekanadham, and Huber.

How will they safely agree on this?[25] When we make the switch to computer science, the generals can be replaced by computers and the messengers are digital communication messages. We achieve "Byzantine fault tolerance" if we are able to solve the problem. Until this point the more "general" explanation (no pun intended). What follows next is the more technical and in-depth explanation. The first sections cover the proof of the Byzantine Generals Problem while the second part goes into the several solutions that have been given for allegory.

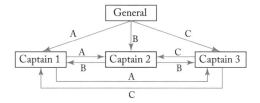

Figure 1.10: The Byzantine Generals Problem.

Several scenarios can be devised in which this allegory can take place. First of all, there are the deterministic protocols with a fixed number of messages. The problem here can quite quickly be spotted. Because we have a fixed number of messages, we can assume that some will be delivered while others are not. The last message received by the receiver will probably not be the last message send by the sender. The other factor that is in play here, is that we are using a deterministic protocol. Therefore, the sender will stick with his original decision. The receiver, on the other hand, will not follow this decision because the final message was not received. We end up with one attacker and one holding off, and the overall attack fails. The second possibility is given by non-deterministic and variable-length protocols.[26] Here we will have to imagine a tree, with its roots all the possible starting messages that can be send out. The branches of the tree are all the following messages with the leaves, or nodes, with all the possible ending messages. We also have the so-called "null trees" that are protocols that end before sending any messages at all. So far so good? Okay, now suppose that we could discover some kind of non-deterministic protocol that solves the problem, by a similar argument as before in the deterministic example, one could derive from the non-deterministic protocol a deterministic one by removing all the ending messages or leave nodes. We are left with a deterministic protocol to solve the problem. We also know that the non-deterministic protocol is finite. We therefore know that the solution would be a null tree. There is only one possible conclusion: a non-deterministic protocol that solves the problem simply doesn't exist. Before the paper

---

[25]   One can imagine sending an infinite number of messages and messengers or just gamble on the chance that one of a number of messengers might succeed. This is of course not the goal of the exercise.

[26]   More detail can be found in *Thought Experiments: Popular Thought Experiments in Philosophy, Physics, Ethics, Computer Science and Mathematics*, by Kennard, F. (2015), Lulu Press, Morrisville, NC.

from Satoshi Nakamoto, several solutions already were coined for this specific problem.[27] One of these early solutions is based on simple mathematics. Instead of looking at the messages being sent, one takes into account the number of generals. As long as the number of traitors among the generals is not equal or less than one third of the total number of generals, we still achieve Byzantine fault tolerance. It starts at a number of three: one general and two lieutenants. If the general sends conflicting messages to the two lieutenants, and they have to check with each other before attacking, one can easily see that none of them can find out who the traitor is. If we increase the numbers, we come to the following formula: the number of generals (nodes) $n > 3^*$ the number of traitors t. You could also try to solve the problem by working with unforgeable message signatures. By making use of public key cryptography one can try to achieve Byzantine fault tolerance because one can always verify the true sender of a message and the message can only be decrypted by the true receiver. A message might still not arrive but when sending a finite number of messages, some will arrive showing the true sender and receiver. A traitor can eventually be identified based on identity and message information. The problem here is that it is not a good solution for safety-critical systems.[28] Several technical implementations were developed over time.[29] A more comprehensive approach was developed by Miguel Castro and Barbara Liskov called "Practical Byzantine Fault Tolerance," or PBFT in short. Also, this protocol was introduced in many different technical implementations, either addressing robustness (such as Aardvark) while others focus on performance (such as Q/U).

1.11.1 DOUBLE SPEND PROBLEM

Double spending is the problem of spending a digital currency more than once. One can easily imagine that this is impossible with physical money, while with digital money, something that one can't hold in their hand and barely see on a screen, information can easily be reproduced. The classic way of preventing double spending is by making use of a centralized authority that, as a trusted third party, is responsible for accepting transactions and the creation of new currency. This way one can verify that no double spending has occurred. One can see several problems with this solution: if one values his or her privacy, a third party might not be the best solution. Certainly because you have to trust this third party to do as intended. The control over the network is also performed by this third party, thereby leaving the other nodes in the network as mere "participants in payments." The most important issue that one could identify is that there is a single point of failure within the

---

[27]   For more details on these solutions, refer to *The Byzantine Generals Problem*, by Lamport, L., Shostak, R., and Pease, M. (1982), ACM Transaction on Programming Language and Systems, 4(3).

[28]   Security vs. safety critical systems. While the first focuses on the intelligence, the second focuses on the mission or action itself. Safety wants to make sure that the action itself occurs as it should while security will focus on keeping the messages themselves hidden. Based on this logic one might think that these enhance each other but this is not always the case. Security measures might undermine safety and vice versa. This is certainly the case for public key encryption.

[29]   Hopkins et al. (1984). *The Evolution of Fault Tolerant Computing*, Springer

network: the central authority. The central power could be compromised or deceived after which the double spending could take place once again. Cryptocurrencies resolve the double-spending problem in a decentralized manner. This way the single point of failure can be removed from the network and one can focus on the participants themselves. To solve this issue there have been created several consensus algorithms which we can divide in two main groups: the proof of work mechanisms and the proof of stake mechanisms. Each deal on their own way with consensus and the double-spending problem.

1.11.2   CAP THEOREM

The CAP theorem or Brewer's theorem (named after Eric Brewer) is a theory that has its foundations in computer science and refers to an impossibility that exists in distributed data stores (such as blockchains). It comes down to the fact that when one is working with a distributed data store, you can only provide two out of three of the following:

- consistency (every read consists out of the most recent write / error);

- availability (every request receives a response); and

- partition tolerance (the network continues to operate despite a number of messages being dropped).

Network partitioning is one of the essences of a distributed system (network failures and nodes falling out) thus this seems to be implied when we want to set up a network, which leaves the choice between consistency and availability. The blockchain implementation opts for partition tolerance and availability with a lesser focus on the actual consistency. The consistency is being supported by making use of a consensus algorithm such as proof of work but a transaction can only be called "confirmed" not after it has been mined in a block, but after several others have followed so that the transaction is not being contested on a later moment.[30]

1.11.3   SPECTRE AND THE CONDORCET PARADOX

This paradox is related to specific to the SPECTRE-protocol (explained later) and similar protocols implemented in blockDAG structures. If this first sentence looks like something from a science-fiction novel it might be better to read on and return when you have read both the explanations on SPECTRE and blockDAGs. The Condorcet paradox (or voting paradox) was created by the Marquis de Condorcet and is in essence a social choice theory. The paradox can be found in the fact that voters don't have cyclic views while this might occur in groups. Easily explained: there are

---

[30]   Nelaturi, K. (February 5, 2018). Understanding blockchain tech – CAP theorem. Mangosearch.com. https://www.mangoresearch.co/understanding-blockchain-tech-cap-theorem/. Accessed June 27, 2019

three candidates we have to vote on. When we look at the group preference, we see that there is a preference of A over B, B over C, and C over A. It is clear that this wouldn't make sense from an individual standpoint but within groups we can find different majorities, made up out of different individuals. This also means that there cannot be no Condorcet winner. Each candidate finds himself in the same symmetrical situation and only if one of the candidates were to leave the election, a majority could be found to elect a winner out of the remaining two candidates.

1.11.4 FERMAT'S LITTLE THEOREM

Fermat's little theorem isn't really a thought experiment but rather a mathematical theory regarding prime numbers. It states that when there is a prime number p, then for any integer a, the number $a^p - a$ is equal to a multiple of p. You can also state this as $a^p = a \pmod p$.

1.11.5 PEDERSEN COMMITMENTS

In several blockchain technologies one makes use of Pedersen commitments. This is used to increase privacy but also to introduce implementations that are able to reduce overhead. We will find this in several platforms such as Monero, Bitcoin, and Ethereum. But what is it? The story begins with a sender that wants to send a secret m in some public messaging space with at least two elements. The second element is a random secret r. The sender has to combine m and r to produce a commitment c by making use of a commitment algorithm C so that $c = C(m,r)$. Next, c is made public and later also m and r are revealed. It is up to the receiver (or verifier) to check that the combination of the commitment algorithm with m and r really leads to the commitment c. In more detail, Pedersen commitments make use of a public group $(G, \cdot)$ of a large order q in which the discrete algorithm is hard and with two random public generators g and h. The random secret r is chosen in $\mathbf{Z}_q$ and the actual secret m is a subset of that. This leads to the following equations:

$$C(m, r) = g^m \cdot h^r$$

It is of crucial importance that the commitment c does not give any indication of m before m is actually revealed. Another condition is that the commitment algorithm should lead to a different outcome when m or r is changed. Also, a different combination of m and r shouldn't lead to the same result. All of this is necessary to prevent specific attacks within the blockchain world (such as double spending).

1.11.6 FUNGIBILITY AND LIVENESS

When we talk about blockchain and distributed networks, fungibility and liveness are some core concepts that you need to understand as these are crucial success factors that need to be addressed for the future of any blockchain platform. Fungibility simply means that something that you use can easily be replaced by something else that has the same and identical function. We generally expect for any currency to be fungible. If I loan someone 20 euro, I really don't care if that person gives me back 4 bills of 5 euro, 1 bill of 20 euro, or any other combination. The value is the same. With this we have immediately landed on another key concept that is tied to fungibility: divisibility. If I have to pay 15 euro, and I pay with a 20 euro bill, I expect 5 euro back. The same rules apply for cryptocurrencies. If we want the general population to make use of cryptocurrencies, we also want to be able to receive change and when I am being paid back, I don't care how many outputs are used to create my input, as long as I get my money. However, there are also non-fungible tokens in place. These are indivisible, cannot be interchanged and have unique properties. What can be the purpose? Well, you are able to create unique assets with rich metadata so that these assets can be traded among participants. These non-fungible tokens cannot be used as a currency but have other interesting applications. There was the cryptokitties craze in 2017. These had unique properties and could (and still can) be traded amongst the participants. A more serious implementation of non-fungible tokens is the creation of unique proofs of identity and unique digital certificates. This can be used for property deeds, academic qualifications, voting (and eliminate election fraud), licensing, and the management of the exchange of personal data.[31] Another concept that is connected with distributed networks and blockchain is "liveness." This idea is strongly connected with "safety" as a consensus algorithm in a network can never guarantee both safety and liveness. Consensus in a network is reached by the nodes that exchange messages and eventually these nodes have to reach a final state. Safety is the guarantee that nothing bad will happen during the search for consensus, while liveness guarantees that eventually something good will happen. Each network has chosen either of safety or liveness as their main priority. The Bitcoin network with Nakamoto consensus has chosen for liveness over safety. An example for an implementation that emphasizes safety is Tendermint that makes use of Byzantine Fault Tolerance-style consensus algorithm to achieve consensus. Problem is that BFT-style consensus protocols in the worst case never achieve consensus as the voting on a block could keep on going on. HotStuff is a protocol that was announced early 2018 that wants to deal with the liveness problem of BFT-style algorithms. This is done by creating blocks that contain the validators votes (or commit-certificates). Blocks without votes can also be created with this protocol, with the risk that finality isn't guaranteed. These blocks reach consensus when other blocks with votes have their finality guaranteed. This is the sacrifice of safety for liveness.[32] This protocol is proposed by Facebook as the consensus protocol for its LIBRA coin (they call it LibraBFT, but it is a protocol that has been derived from HotStuff).

[31] Chandraker, A., Kachhela, J., and Wright, A. (2019) Digital identity, cats and why fungibility is key to blockchain's future, *PA Consulting*. https://www.paconsulting.com/insights/blockchain-fungibility-future/. Accessed June 26, 2019

[32] Woo Kim, S. (May 28, 2018) Safety and liveness – Blockchain in the point of view of FLP impossibility. *Medium*. https://medium.com/codechain/safety-and-liveness-blockchain-in-the-point-of-view-of-flp-impossibility-182e33927ce6. Accessed June 28, 2019

1.11.7 TRANSACTION AND SETTLEMENT FINALITY

Transaction (or operational) finality is another major concept that one needs to understand when they want to discuss the future and acceptance of blockchain and cryptocurrencies. Finality is a concept well-known in the financial world and probably with all of you. It is the general understanding that when you perform an operation, that operation is completed for good and doesn't change in some moment in the future. You also have settlement finality which is a statutory, regulatory and contractual construct in which you agree on a moment in time when a party has discharged an obligation or to have transferred an asset or financial instrument to another party, and this becomes unconditional and irrevocable despite insolvency or bankruptcy.[33] Based on the nature of the blockchain network, transaction finality can be provided, is probabilistic, or can't be provided at all, which in turn influences how the network can actually be used. You can understand that when it comes to financial transactions this is quite important to the participants. If I can know in a quick and concise manner that a transaction is final, I can provide goods and services for this transaction. But what if you weren't able to get this certainty? Would you still be willing to perform services for that transaction? No, of course you wouldn't. But are the current financial systems we rely on that secure? Well, Vitalik Buterin argues that you can never be 100% certain that a transaction is final.[34] There could be corruption at the central bank, systems can fail, hackers can change information, paper ledgers can be stolen or burn, and I am sure that you can think of some other examples. This means that you rely on probability when it comes to the concept of finality. In blockchain networks, one can imagine attackers taking certain assets or cryptocurrencies from other participants. As it is with any case of theft, it is up to the court system to determine the real ownership of a certain asset when it comes to theft or malicious intent. In the past, there have been some reversals in blockchain networks, which leads to the reversal of other transactions or to splits that have to be resolved over time. So, when do you accept a transaction as final when you look at a decentralized network? Well, when several nodes start accepting the transaction and want to mine it. So, if you really want to be certain that a transaction is valid, run multiple nodes and look if they all accept the transaction (whether you are making use of a public chain or a consortium is beside the point in this case). When we look at proof of work blockchains, a transaction is never truly finalized as one can create a chain with more mining power and overcome the main chain. However, in general six confirmations are accepted for a transaction to be final. If you assume that an attacker has less than 25% of the hashing power in a network, there is a probability of an attacker succeeding to overcome your transaction of 0.00137. If you wait 13 transactions, this is even reduced to 1 in a million.

[33] Liao, N. (June 9, 2017). On settlement finality and distributed ledger technology. *Yale Journal on Regulation*. yalejreg.com/nc/on-settlement-finality-and-distributed-ledger-technology-by-nancy-liao/. Accessed June 30, 2019

[34] Buterin, V. (May 9, 2016). On settlement finality. *Ethereum blog*. https://blog.ethereum.org/2016/05/09/on-settlement-finality/. Accessed July 2, 2019.

However, one can still imagine certain attacks on the network such as the P + epsilon attack or the Maginot line attack (explained later). Proof of stake protocols can offer even more security when they make use of predefined voters that have to put their stake behind their vote. Voters voting the other way, and losing, lose their entire stake. This means that a transaction can still be reverted but only at enormous cost of the voters, leading to an incentive to act honestly. A final point is that, in a proof of stake network, a participant cannot be forced to follow a certain chain, but can choose in case of a fork which transactions he believes to be true. As you can see, transaction finality can be guaranteed to a certain point in a distributed network, similar to what you expect from a central authority. Settlement finality is something different. The "problem" is that settlement finality refers to a point in time while blockchain networks reach consensus over time. Settlement finality is in the end a concept that is clearly defined by legal frameworks and as it is a legal construct, this differs from jurisdiction to jurisdiction. This means that some jurisdictions and countries can easily accept the probabilistic finality that is offered by a distributed network while others will see a blockchain network as never truly "final." Important is that permissioned networks will find acceptance much quicker within the existing frameworks as it relates to a defined number of participants that can interact with each other and exchange, i.e., fiat currencies. However, permissionless ledgers are something novel when it comes to legal frameworks and with the current definitions, it might be more difficult to reach settlement finality. This because such networks operate cross-border, make use of cryptocurrencies (which have properties of both fiat currencies and assets) and the current legal framework isn't prepared for the issuance and transferring of cryptocurrencies, leading to uncertainty when we talk about settlement finality. However, this doesn't mean that it will never happen. The technology is ready, now we just need the legal frameworks to support it.

1.11.8 CENSORSHIP RESISTANCE

When you read up on public blockchain platforms, you will almost certainly run into the concept of censorship resistance, something that cannot be offered by any permissioned blockchain network. So, what is it? Censorship resistance means that anyone can transact with the network on the same terms, regardless of their identity, status, or any other criteria. The only rules that need to be adhered to are the rules of the network. You can imagine that networks that completely secure the privacy of their participants, are more censorship resistant than others. However, there is always a "but." Public ledgers are also able to exclude participants of making use of certain decentralized applications that are built on top of them. This seems to be counter to what you would like to hear. The reason why Bitcoin became popular in the first place was the fact that it was censorship resistant. Even to this day it is very popular all over the world. In countries where the fiat currency experiences hyperinflation or where people live under international sanctions, such as Venezuela and Iran, Bitcoin offers a means of payment (and even stability to a sense). Again, there is a "but." To increase acceptance of public networks, decentralized applications on top of them can be blocked for certain participants. There is the case of the Tron network that works together with the Japanese government to prevent citizens to excessively use gambling applications.[35] This introduces censorship in an otherwise open network! Also Ethereum allows decentralized applications to exclude participants, as, i.e., financial institutions have to be able to exclude participants based on certain criteria, and this way they can also create applications on top of the network. So public blockchain platforms can be censorship resistant, to a point.

1.11.9 WINTERNITZ ONE-TIME SIGNATURES

Winternitz one-time signatures or "W-OTS" are used in several schemes within blockchain networks to increase security and prepare for the age of quantum computing. An example of a network that has already implemented the scheme is IOTA. In 1979 the concept of one-time signature was introduced by Leslie Lamport. It relied completely on hash functions for the security proof. You could use these once to keep the security of your messaging scheme (public–private).

Your public key looks like: $h(x0)|h(y0)|h(x1)|h(y1)|\ldots$

Your private key looks like: $(x0, y0, x1, y1, \ldots)$.

After only a few months, Robert Winternitz proposed a different scheme where $h^w(x)$ is used instead of $h(x)|h(y)$. There is also a short checksum added to prevent attackers from guessing the secret x when they figure out w.

1.12   WHAT IS BLOCKCHAIN?

1.12.1 FOUNDATIONS

When we talk about blockchain, we need to talk about the foundations. If you already have a general understanding of blockchain and distributed ledger technology, feel free to skip this part. If not, I will try to remain as general as possible in these sections, as deeper explanations will follow later on in this book. When we talk about blockchain, we talk about ledgers. In the classic way of working there is a third-party necessary to verify payments, notarize transactions, make use of escrow, and allow voting and registration. With the advent of blockchains, we move away from these third parties which centralize power. In itself, centralization isn't necessarily bad but it often leads to inefficiency, high cost, loss of privacy and control, corruption, and more. With a distributed network this is no longer possible, as the participants decide the outcome and validity of transactions. There is a single truth that must be supported by the majority, and not by a single power in the network. People often

---

[35] Sedgwick, K. (April 4, 2019). Decentralized networks aren't as censorship-resistant as you think. *News.Bitcoin. com*. https://news.bitcoin.com/decentralized-networks-arent-as-censorship-resistant-as-you-think/. Accessed July 2, 2019.

forget what this technology actually is, and instantly start looking at the implementations. But to understand the possibilities, you first need to truly know and realize what this technology represents. One important part is the existence of a P2P network, which we explained before. We remove any centralized servers and allow the participant nodes to directly communicate with each other. A second important technology to understand is asymmetric cryptography. It basically exists out of two keys: a public key and a private key. As you could guess, the public key can be shared with the public, and the private key should remain private. In many blockchain implementations, the public key is better known as the "account" address where you can send cryptocurrencies such as bitcoin or ether to. The private key is the "password" to the account. The final part to understand the technology is the concept of hashing. When you hash a certain dataset, you receive a unique identifier of that dataset. If you only change one number, letter, or symbol, the resulting hash will look completely different. And when I say completely, I mean COMPLETELY. This way you have an assurance that the data has not been modified. As you can see below, the difference between the hashes of the word "apple'" and the word "apples" is clear.

In blockchain this technique is used so that resulting datasets can rather fast be compared with each other. It uses a technique called the "Merkle tree" (details above), which allows to create a hashing structure based on other hashes, which again refer to datasets (transactions). These are used afterwards in the mined blocks. All of these blocks are sets of transactions that are chained together in one huge ledger. The blockchain, in fact, represents nothing more than a historical overview of the transactions that have taken place in the network. All these techniques together help us create an append-only ledger of transactions which is distributed over all the nodes in the network. These transactions are contained in a linked list of blocks.

1.12.2 BUILDING "BLOCKS"

The term "blockchain" refers to the very core of the technology.[36] The data structure underlying is an ordered back-linked list of blocks which consist out of transactions. These transactions are exactly what you would think they are. They represent payments between all the participants taking part

[36] Van Hijfte, S. (2020). *Decoding Blockchain for Business.* 1st ed. New York: Apress.

in the blockchain network. You could store this data in several ways, as a database or even as a flat file. The way the blocks are linked to each other, is by making use of the hash of the previous block that was last added to the chain. If you think about it, the blockchain with the transaction data is nothing more than a ledger where the data has been structured in a new way.

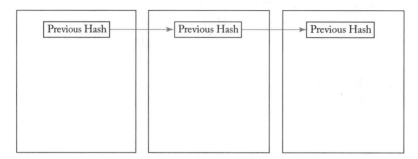

Figure 1.11: Linking together of blocks.

This hash is used in the calculation of the hash of the next block, therefore linking this information to the core of the system. Because we know that each hash is a unique fingerprint, we can always know for sure that the block is part of the chain (i.e., when we try to look up a certain transaction with a block explorer). Important to note here is that a parent block (the most recently added block) can have more than one child blocks, while a child will always have one parent. When there is more than one child, we are dealing with a fork in the system. Normally these forks will be resolved and only one of the children is used to continue the chain. In some cases, however, as we will see later, some of these forks stay on and we will deal with separate chains that all once had the same parent. We always need to have a starting point, a first block, which is conveniently called the "genesis block." Because we could visually understand that blocks are being stacked upon each other, we refer to the most recent block as the "tip" or "top" and the distance to the genesis block as the "height."[37] The greater this "height" becomes, the more difficult it will be to make a change to one of these earlier blocks. The longer the chain becomes from a certain block, the more computer power it will require to recalculate the information contained in all the blocks. This is also immediately the most important security that is offered by the technology. As we mentioned before, the blocks are linked to each other by making use of the hash of the previous block. But what else is stored in the blocks? We have two main parts between which we have to distinguish: the block itself and the block header. The hash of the previous block is one of the items that is stored in the block header. That is not the only thing though, depending on the blockchain platform you can also find the Merkle root, timestamp and nonce in there with a couple of other parameters (i.e., difficulty target, version, etc.). Below you can find the example for the Bitcoin blockchain.

[37] One could try to use this height to try to identify a block but this is error prone, as the height is not a unique identifier. The hash, on the other hand, will give you this unique identifier.

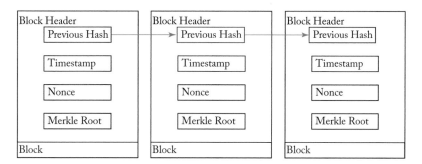

Figure 1.12: Block header information Bitcoin

Not all this information might be clear for now but no worries, we will explain as we go on in the book. Important for now is that you still remember what a Merkle root is. It is this Merkle root that gives us the digital fingerprint of the transactions that are stored in the block itself. As we explained before, the Merkle root is a "hash of hashes" based on the transaction IDs of TXIDs. This hash is unique for the transactions in the block itself.

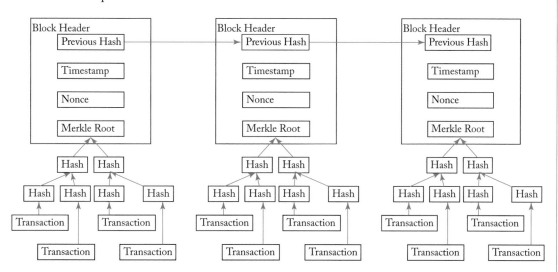

Figure 1.13: What can be found in the Merkle root for Bitcoin.

Again, this is a simplified view of what can be found in the block headers but for now it can give you at least a first understanding of what you can find in a block header. Of course, the block consists out of more than only the block header. The "bulk" of a block is made up out of the transactions themselves.

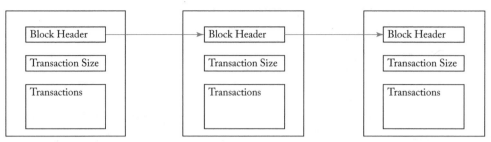

Figure 1.14: The entire chain.

As you can see, there is a lot of data, not only in the blocks, but also in the block header. The header is crucial for creating the link between all the blocks as we contain in this header the hash of the previous block. It is the hash that is ultimately calculated that can correctly and uniquely identify a block within the entire blockchain.[38] For Bitcoin, the SHA-256 algorithm combined with RIPEMD-160 is used to calculate these hashes. Now we have an idea of how transactions are put into blocks and how everything is linking together, the main question remains: how do we perform transactions? The "how" in technical detail will be shown later on in the section, as this differs from one platform to another, but the idea is that you can sign your transactions with your personal key in such a way that it is not only clear that you are the person that are spending it, but also that you have the right to spend it and that you didn't spend the currency before. So if we look back at the story so far, we have a P2P network of nodes which communicate with one another by means of a "gossip" protocol. (We will explain this later on.) Transactions are broadcast through the network and mined into blocks. These transactions represent state transitions that need to be processed by the state machine of the blockchain network. These will only be processed if these transactions are performed based on the consensus rules that all participants agree on. So, in short, the lifecycle of a transaction is as follows: a transaction is created and signed by the creator. This transaction should immediately contain all information necessary to verify and execute the transaction. These proposed transactions are shared with other nodes in the network which verify them. If accepted, they are propagated throughout the network, otherwise they are disgarded. Mining nodes can include these transactions into blocks that can be mined and eventually added to the chain of earlier accepted transactions.

1.12.3 BLOCKCHAIN OR DISTRIBUTED LEDGER?

An important note that I immediately want to make here is the distinction between blockchain and distributed ledger technology (DLT). Often it is said that each blockchain is a distributed ledger but not every distributed ledger is a blockchain. So you can see that there are some simi-

---

[38] The hash of the genesis block within the Bitcoin network is
000000000019d6689c085ae165831e934ff763ae46a2a6c172b3f1b60a8ce26f.

larities in concept and what they try to achieve. Distributed ledgers are databases that try to share data among geographical different locations without a central actor having control over the entire network. A major difference between the two is how new data is appended to the platform. While in blockchain technology one makes use of a consensus algorithm to add new information, distributed ledgers don't always have such an algorithm in place. The application of DLTs differ in many forms as you will see in this book, just as the application of blockchain platforms. The use of name "blockchain" or "distributed ledger" has many implications. Not only is the technology different but also the perception of the two names. While the first is well known and hyped all over the world, the second remains more hidden in the shadows and seems to be more known among IT specialists. So why use one name or the other? Some companies want to ride the hype and use "blockchain" while others want to step away from the hype and show that they are really focusing on the technology itself and use "distributed ledger." Another point one should take into account is the "types" of blockchain that currently exist. There are the public, permissionless blockchains (such as Bitcoin and Ethereum) which have no restrictions when it comes to either access or participation. We can also call them the "true" blockchains. We also have private, permissioned blockchains where only a certain group of people can gain access and participate (such as Rubix and Hyperledger platforms). You also have some platforms that exist in-between these two "extremes." The first are the public but permissioned blockchains which allow everyone to transact and see the transaction log while only a few can participate in the consensus mechanism (such as Ripple and private versions of Ethereum). Finally, there are also the private but permissionless blockchains where the consensus algorithm is open to everyone while transactions are limited to a specific number of participants. There is no real example available of a network that fully achieves this (the one that comes closest currently might be the Exonum network). Next to these examples there are also the directed acyclic graphs (DAGs) that have also entered the blockchain space with their own solutions and networks. To not further complicate the progress in this book, I will use the term blockchain and distributed ledger as synonyms (I know you might not agree with this approach), and will specifically refer to DAGs in case we are discussing them in more detail.

1.12.4 BLOCKCHAIN ADDRESS

The next step in understanding the world of blockchain is the blockchain address. A blockchain address[39] is one of the main concepts within blockchain and cryptocurrencies to understand. It is based on public key cryptography (also known as asymmetric cryptography) where one makes use of a private key and a public key. As you could infer from the name, public keys are keys that can be known by the public. They are mainly used to identify you, while private keys should always be

---

[39]   In the early days of the Bitcoin network, you could pay directly to an IP address. You can imagine several problems with this, such as "man-in-the-middle" attacks. This is why this system was abandoned in favor of more secure options.

kept private. Imagine that I would like to send a message to you, the reader, and I wouldn't want anyone else to read it, and I could make use of public key cryptography. By encrypting a message with your public key, I make sure that you are the only person that is able to read it. You would need your private key to decrypt the message. Similarly, if you would like to reply on the message, you would encrypt it using my public key, after which I would use my private key to decrypt the message. There is a wide arrange of algorithms that is able to create such public and private keys using one-way functions. You should be aware that public key cryptography has a lot of uses in computer science.[40] The same way public key cryptography can be used in blockchain networks. If payments are being sent over the network to blockchain addresses (which are hashed public keys) these can only be unlocked by making use of private keys. In the Bitcoin network, for example,[41] you have several possible addresses which can be used.[42] One of these is the P2PKH address, also known as the "Pay to public key hash" address. The wallet creates an address by using the elliptic curve digital signature algorithm (ECDSA), and entropy to generate the private key. From this private key the public key is derived. This public key is then hashed[43] with SHA256 and afterward with RIPEMD-160. Afterward, the prefix bytes of "00" is added to the result which is the reason why a Bitcoin address starts with a "1" and why there are 4 checksum bytes on the end. These checksum bytes are generated by double SHA256 hashing and taking the first 4 bytes of the result. Finally, the result is converted into a base58 string. A more advanced example is the "P2SH address" or "pay to script hash" address.[44] Instead of paying to the hash of a public key, you pay to the hash of a script. Here you need to provide a script to unlock the account linked to the hash. These addresses add the byte prefix "05" and this is the reason why they start with a "3." Finally, you also have the "P2WPKH" or "Pay to witness public key hash" in the Bitcoin network. This was introduced with segregated witness.[45] Based on the manner that these addresses are derived, it can be possible to store altcoins on the same address. Coins such as Litecoin, Dash and Dogecoin are all derived in a similar manner as Bitcoin. The main difference is the prefix that is added. They each use their own prefix to identify the altcoin ("D" for Dogecoin, "X" for Dash, and "L" for Litecoin). This means that you can use the same public and private keys to store these coins. Litecoin uses the same prefix for P2SH addresses and this way you can store the coin on the same address. However, you could identify a whole number of cryptocurrencies which use different algorithms to calculate the address, e.g., Monero,[46] where the cryptonote

[40]   Think about PGP, S/MIME, GPG, SSH, SRTP, etc.

[41]   Hoogendoorn, R. (December 3, 2019). Easypaysy makes Bitcoin addresses much easier. *Medium*. https://medium.com/@nederob/easypaysy-makes-bitcoin-addresses-much-easier-faf40988614. Accessed June 4, 2020.

[42]   P2PK is where you would send funds directly to the public key without concealing it with an hash. This seems to be inherently unsafe.

[43]   This to increase the security with quantum computing on the rise.

[44]   https://bitcoin.stackexchange.com/questions/64733/what-is-p2pk-p2pkh-p2sh-p2wpkh-eli5.

[45]   The scriptSig with its parameter is replaced with "witness" in the transaction to verify the validity.

[46]   Rosic, A. (2017). Blockchain address 101: What are addresses on blockchains? *Blockgeeks*. https://blockgeeks.com/guides/blockchain-address-101/. Accessed July 4, 2019.

algorithm is used. Not only is edDSA used here, there is also the use of "ring signatures" to provide more privacy and therefore you need two public keys, a view and a spend key. It also uses Keccak-256 instead of a double SHA256. Because it is quite hard to remember these addresses, several alternatives have been thought of and a more recent one for Bitcoin is called Easypaysy where Bitcoin addresses are transformed to something that looks like an email address.

Ethereum takes another approach to the generation of blockchain addresses. Here you also start with a private key and ECDSA to derive the 64-byte public key. Next, the public key is hashed with Keccak-256 which results in a 32-byte string. The first 12 bytes are dropped and to the remaining 20 bytes, the prefix "0x" is added. This doesn't seem very secure if you compare it to address generation on other blockchain networks. In the beginning, Ethereum developers didn't really care as much about the security because their main focus was the development and possibilities of the platform. There is also another reason why this isn't as important: by making use of smart contracts, one can adapt and transform the addresses as needed, there will always be a reference to the original address. The ICAP-format, or "Inter Exchange Client Address Protocol" is gaining more and more support in the community.[47] This new design uses the IBAN format which is widely used in banks (or BBAN in the UK).[48] In the ICAP-format, the country code is replaced by "XE" after which the BBAN can be split in three possibilities: direct, basic, or indirect. The BBAN for the code when "direct" will consist out of 30 characters and this will be interpreted as a big-endian encoded base-36 integer representing the least significant bits of an Ethereum address. The "basic" option will be non-compliant for IBAN as it will consist out of 31 characters instead of the 30 characters mentioned before in the "direct" option. Finally, there is the "indirect" option and here there will be 16 characters and will comprise out of 3 fields: an asset identifier of 3 characters, institution identifier of 4 characters, and institution client identifier of 9 characters.

1.12.5 BLOCKCHAIN WALLET

Closely related to addresses is the concept of blockchain wallets. These wallets aren't used to store cryptocurrencies but are used to interact with the network. They are used to generate the information necessary to send and receive cryptocurrency and to do this they make use of public and private keys. The public key, or address (as explained before), is used to receive transactions. The private key is used to sign transactions.[49] Depending on how you make use of a wallet, they can be defined as either "cold" or "hot." A hot wallet is the easiest to understand, as it is a wallet that is connected to the internet. There are several providers out there that will allow you to make a

---

[47] Chen, M. (April 13, 2019). Inter exchange client address protocol (ICAP). *Github - Ethereum*. https://github. com/ethereum/wiki/wiki/Inter-exchange-Client-Address-Protocol-(ICAP). Accessed July 3, 2019.

[48] The IBAN contains three pieces of information: the country code, error detection code, and the basic bank account number.

[49] Or seed phrase, depending on the wallet you are using.

wallet. Hot wallets are also called software wallets and they come in several different kinds. There are the web wallets which can be created in a browser, another type is the desktop wallet. This can be downloaded on your machine and are therefore considered safer than web wallets. Still, you will have to keep your wallet safe and take backups if possible. Another type of software wallet is the mobile wallet, which is an implementation specifically for your smartphone. Again, backups seem to be necessary to make sure you don't lose your wallet with your phone and encrypt the phone to prevent cyberattacks or at least mitigate them to a certain level. A cold wallet, on the other hand, has no connection to the internet, and is used to store cryptocurrencies offline. This is a much safer way of storing (keeping in mind you don't lose your cold wallet) as hot wallets can be prone to cyberattacks. A hardware wallet is a first form of cold wallet. These are physical devices that are used to store tokens for a longer time. There are also implementations that can be used similar to perform transactions. The problem here can be the firmware implementation of the wallet, which is not always as secure as it should be. A smartphone permanently kept offline can be seen as a hardware wallet with similar security. Another interesting example is the ZERO developed by NGRAVE which is a hardware, cold wallet with extreme security measures in place. Finally, there are also paper wallets. As you might have imagined, this is simply a piece of paper with QR codes that contain the public and private keys. Paper wallets are very dangerous, as a piece of paper is clearly open to specific dangers. On top of that, these types of wallets can only be used once—to send the entire amount to another address.

1.12.6 NODE

When we talk about blockchain and the network it represents, we also speak about nodes. Nodes are the lifeblood of the network as they are always responsible for a given set of tasks. Without the nodes, the network would no longer exist, even if the software would still be up to date. These nodes are distributed all over the world, across a widespread network. So, what is a node? A node is any electronic device that is connected to the network and has an IP address. One of the main purposes of the network is to maintain a copy of the blockchain and process transactions (depending on the type of node). The owners of these nodes willingly use their hardware, computer power, and energy to maintain the network. To be rewarded for this, miners have the chance to collect a reward, based on the transaction fees within a block and new coins being minted. This is known as mining or forging. Depending on the type of blockchain network, this can require huge amounts of computer power and linked with this, a huge cost of electricity. We should also make a distinction between two types of nodes. It can either be a communication endpoint or a point of communication re-distribution. Even though they are equal throughout the network, each type supports the network in a different manner. First of all, there is the full node which downloads a complete copy of the blockchain and checks for any new transactions based on the consensus protocol that is in use. A light node on the other hand is referencing the copy of the blockchain on a full node.

The network nodes have a tree-like structure so that we can also identify the following terms:

- **Root node:** this is the highest node in a binary tree;

- **Leaf node:** a node with no children;

- **Tree:** a structure of nodes;

- **Forest:** a set of tree;

- **Parent node:** node that has other nodes extending;

- **Child node:** linked to a parent;

- **Sibling node:** node connected to the same parent;

- **Edge:** the connection between the nodes; and

- **Degree:** the number of children of the node

It is important to know and understand that the security and strength of a blockchain network originates in a large extent from the number of nodes. The more nodes in the network, the more distributed the power is in the network and the lower the chance that a malicious actor will try to take advantage of the network.

1.12.7 MINING

Mining is one of the key concepts within blockchain technology. It is the way new transactions are being accepted within new blocks[50] that are added to the existing chain, as well as how new currency is being created. It is always used as a countermeasure against fraud and makes sure that all participants within the network remain true. The mining itself is quite costly, as it requires hardware to be used for the mining process, energy to power the mining itself, and time. For this, the miner should be rewarded and this is done in two ways: the miner receives the transaction fees of the transactions that are included in the block and the new coins that are being created when a new block is added. The miner can receive this reward based on the algorithm that is being used within the network, either proof of work or proof of stake. The mining process is not only the key to the creation of new cryptocurrency, it is also the mechanism that helps create decentralized consensus in a trustless environment. All nodes over the network receive the blocks and can consequently check its validity. This means that consensus is will emerge over time as there is not an election at

---

[50]   In the case of Bitcoin, a new block is being added every 10 min.

a specific time but by an asynchronous interaction of all the nodes in the network. You must real-ize that in networks such as Bitcoin, the computer power necessary to compete and mine for the next block has increased exponentially over time. This is because of the increase of entrants in the market space but also because of evolutions in hardware solutions.[51] Over time, mining pools saw the light of day. By working together, the pool has a higher chance to find the next winning block so that the rewards can be shared among the participants. The infrastructure of mining installations has evolved greatly over time and, depending on the blockchain platform, we will go deeper into the when and why of these developments. For some of these platforms one needs quite advanced infrastructure, while others are still open for everyone. So what infrastructure is actually necessary to perform mining? Depending on the network and consensus protocol in place, the least you need is a CPU (central processing unit). In 2009 and early 2010 this was still in use for the Bitcoin network, but was later replaced with GPU (graphical processing unit) mining. These GPUs were eventually replaced at the end of 2011 by FGPA (field programmable gate array) mining. Eventually, in 2013 the rise of ASIC (application-specific integrated circuits) made sure that network-specific mining infrastructure could be created. This is very expensive mining hardware and in the case of Bitcoin, this even gave rise to specific Bitcoin mining startups. This immediately leads to a lot of criticism. First of all, it makes sure that the network no longer is really distributed and accessible to everyone when it comes to earning cryptocurrency and power comes in the hand of mighty corporations who do have the cash to buy such infrastructures. Also, the power consumption is enormous, certainly when you know that a network such as Bitcoin, on itself, already consumes the power of a small country! Certainly if the network wants to keep the same block time, this increase in participants willing to mine, leads to more difficult computations, which leads to more need for computing power (an endless cycle). On top of that, the cost for transactions also increases, making the net-work even less desirable for day-to-day use. This has led to several reactions by other networks, such as changing to other consensus protocols, e.g., proof of stake. Here the energy consumption is much lower and the mining is based on reputation. The stake might still be high but can be adjusted based on the needs of the network participants. Another reaction has been to implement alternative proof of work consensus protocols which are ASIC resistant. This means it is very difficult to build a specific mining infrastructure if these protocols are in place, so that participants have to rely on, i.e., FPGA mining which is open for a lot more participants.

1.12.8 MERGE MINING

Merge mining is a special case where the miner is actually capable of mining more than one chain.[52] This does not mean that the chains have to be related in any way or that they have to contain data

---

[51]   There was the shift from CPU to GPU and FPGA mining. Later there was the introduction of ASIC mining.

[52]   Roberts, D. (January 9, 2014). Mergen-Mining.mediawiki. *Github – Namecoin*. https://github.com/namecoin/wiki/blob/master/Merged-Mining.mediawiki. Accessed July 6, 2019.

of one another. You once again create hashes based on the transactions you are trying to mine. If you finally find a solution, it is up to you to provide the solution to both chains. If it is correct for the first chain, you receive tokens of this chain, if it is correct for the second chain you subsequently receive tokens of the second chain. If it is correct for none, you simply receive nothing. The extra security that is being built into the system is because of the link that can be created between 2 or more chains. An example is the relation between Bitcoin and Namecoin. When a Namecoin block is created, this is hashed and included in a Bitcoin block as a transaction hash, therefore effectively linking the Namecoin block to a Bitcoin block. While, on the other hand, in a Namecoin block you can also find a Bitcoin block header. This way the Bitcoin chain is linked to the Namecoin chain. This block header is used to as a proof of work. It clearly shows that only the mining is linked, improving the security (against for example a 51% attack). [53]

1.13   BLOCK TIME

Block time is an important concept that one needs to understand. Vitalik Buterin has published several blog posts on the concept and why it is important. Bitcoin has a block time of 10 min on average while the Ethereum network has a block time that varies between 12 and 17 sec (the white paper states 12 seconds while the reality, due to a specific difficulty, is around 17 sec). The block time, as you might have guessed, is the time necessary for a new block to be accepted on the blockchain. Each network has its own time because of the choices that were made by the creators. A long block time can create frustration for its users. Certainly, if sellers are careful for certain attacks such as the Finney attack or the double-spend attack (explained later), it can be better to wait a couple of blocks before the transaction is actually accepted. [54] So why not immediately go for a high block acceptance speed? Well, here is the problem: blocks are mined based on the score that they are given. This score depends on the distance between a certain block and the genesis, or first block, in the chain. The block with the highest score is seen as the correct one and mined. But there is a problem, stated by Decker and Wattenhoffer in their paper "Information Propagation in the Bitcoin Network" where they state that it takes 6.5 sec for a block to reach 50% of the nodes and 40 sec to reach 95% of the nodes. At a rate of 10 min per block there isn't a problem but at a rate of 12 sec per block, you can easily understand that the frequency of miners finding a block only seconds apart from each other increases. The first miner always wins, leading to more stale blocks in the network. This again leads to insecurities such as the 51% attack that no longer requires 51% of the network and the problem of mining centralization (there is an efficiency gain for mining

[53]   Schwartz, D. (August 31, 2011). How does merged mining work? *Stackexchange*, https://bitcoin.stackexchange. com/questions/273/how-does-merged-mining-work. Accessed July 10, 2019.

[54]   Buterin, V. (July 11, 2014). Toward a 12-second block time. *Ethereum Blog*. https://blog.ethereum. org/2014/07/11/toward-a-12-second-block-time/#:~:text=At%2012%20seconds%20per%20block,a%20 stale%20rate%20of%2050%25. Accessed July 11, 2019.

pools vs. single miners in this situation, leading to centralization, the very thing one would want to avoid in a public, permissionless blockchain ecosystem). Decker and Wattenhoffer did offer some possible solutions to speed up the transmission time of the blocks through the network, such as broadcasting the header first, and only the block itself after or simply cutting the block size (because the transmission time is linked to the size of the block). Another blog post by Vitalik Buterin ("On Slow and Fast Block Times") concludes that faster block times can be beneficial because they provide granularity of information throughout the network. In case of a fork, the network can quicker decide on the right path to continue on. He also clearly states that it is a balancing act between user experience, scalability, and usability vs. security concerns such as centralization risk and higher stale rates. The trade-off between slower or faster is not perfect and completely relies on the mechanisms that are built into the network. You will soon learn that each platform has made their own choices based on whether they are private or public, permissioned or permissionless.

1.14   THE CONSENSUS PROTOCOL

We quite often talk about protocols when we talk about blockchain technology. This is of course not something that is only limited to blockchain but can be found in any implementation of telecommunication technology. When we are talking about a protocol, we are talking about an entire set of rules which decide how you can connect to a system and interact with it. These rules can be really extensive as they can determine which hardware you have to use, which software is allowed, and what the parameters are of messages that are being transmitted over a network. As with other telecommunication services, this is the case for blockchain. When we talk about open-source block-chain implementations, like Bitcoin or Ethereum, there are no restrictions on hardware and the software needed is completely free. Even though now this is also still the case for private blockchain (or rather distributed ledger) implementations, one could see future developments where this would no longer be the case.

1.15   ROUND ROBIN

More suitable for a private blockchain than a public one, round robin allows the participant adding the block to sign the transaction. This system works when the participants actually know each other and there is a certain level of trust. Within selected time-frames specific participants are allowed to create new blocks which can be added to the chain, making sure that not a single participant can take over the network.

1.16   PROOF OF WORK

Proof of work was the first consensus protocol to be used within a blockchain network. The first network to implement this type of consensus protocol was the Bitcoin network that was afterward used by many other networks. The idea is that miners have to use their nodes to solve a mathematical problem. It will require a lot of work and computer power to solve, but verifying the result should be easy. This way of consensus is designed to be difficult and requires a lot of fire power. A target hash will be set by the network and the nodes have to try to compute a hash based on the block and the nonce that will be below this target number. The lower the target is set, the more difficult it will be for the participants to find a correct and acceptable hash. The proof of work protocol can help to address the issue of Byzantine fault tolerance by making use of the before-mentioned nonce and by combining messages into blocks. To prevent precomputation, the nonce is unique for each node and can only be used once. An important point of criticism on this type of protocol is the amount of energy that is being consumed by networks that apply this type of protocol. In times of climate change, scarce resources, and economic crisis, this is an important point to consider. There are a lot of different proof of work consensus protocols that are currently in use by several networks and several of these will be described here. It is only a limited list as any number of protocols can be created and put into use based on the use case you are working on.

1.17   NAKAMOTO CONSENSUS

Nakamoto consensus is a term you will often see re-appear with blockchain platforms and their consensus protocols. For Nakamoto consensus to take place we need proof of work consensus, block selection, scarcity, and an incentive structure.[55] These are all the rules that govern the Bitcoin network and have been used by a lot of other platforms since. The proof of work algorithm makes sure that computational power is needed to come to the correct consensus and no 51% attacks can take place in the network. These miners are mining a lottery-type of algorithm in the hopes of mining the correct outcome and therefore select the next block for the blockchain. The only way of wining is contributing enough power in the hope of making a good chance to mine the block. The third concept, scarcity, is created by the limited amount of Bitcoin that can be mined. Only 21 million can ever be mined. The final part, the incentive structure, has been set up by rewarding the miners when they mine the winning block. This way the network remains socially scalable while the participants are encouraged to stay honest and invest their time and power to keep up the entire blockchain structure.

---

[55]   Curran, B. (June 26, 2018). What is Nakamoto Consensus? Complete beginner's guide. *Blockonomi*. https://blockonomi.com/nakamoto-consensus/. Accessed July 12, 2019.

1.18   PROOF OF STAKE

Proof of stake was developed after the proof of work protocol and is more and more being used within blockchain networks. The first network to implement this kind of protocol was Peercoin in 2012. In the proof of stake network, the miner of the next block is selected pseudo randomly as the amount of cryptocurrency held by the node influences the chances of being chosen. The probability of being chosen is thus directly linked to the stake you have in the network. It is clearly more cost effective than the proof of work consensus protocol as miners don't have to use energy to solve a mathematical problem. Second, it has proven to be more secure. It helps prevent the 51% attack. This might seems contradictory, but the stakeholders with the highest stakes are motivated to maintain the network, because if an attack would occur this would damage the reputation of the network and hurt these participants as the value of their stakes would diminish. There is also a downside to this protocol, called the "nothing at stake" problem. When there is a consensus failure in the network, and the participants in the network have nothing to lose, there is nothing to stop these participants from supporting different side chains.

1.19   DELEGATED PROOF OF STAKE

The delegated proof of stake protocol maintains an irrefutable agreement on the truth across the network. The protocol makes use of real-time voting combined with reputation to achieve consensus. This allows every holder of cryptocurrency to influence the network. This network makes use of delegates which are elected in their roles and have to put a certain amount of cryptocurrency within a base account. The larger this amount is, the more influence the delegate can exert over the network. In case of malicious behavior, the money in the base account is lost. We can also call this deposit-based proof of stake. While the delegates are responsible for the validation of transaction, it is up to the participants to request regularly if the blocks mined contain all the correct transactions. This makes sure that the network is self-governed and policed. You can immediately sense that this is more democratic than the other consensus protocols.

1.20   PROOF OF AUTHORITY

Proof of authority (PoA) is an alternative that is often used by private blockchain[56] networks (more related toward distributed ledger networks) where proof of work is replaced by the "identity" of the nodes as a stake in the network. Only these selected nodes are allowed to mine new blocks. Only these "validator" nodes are allowed to add transactions to the blocks that are consequently added to the blockchain. With proof of authority and validators, there is also the new concept of "reputation." The reputation of the validators is crucial for the existence of the network. It requires

---

[56]   There are also certain public networks that make use of this protocol.

validators to invest money and to confirm their real identities. This reduces the risk of malicious activity. If the reputation of one of the validators or the "validator authority" is damaged, the other participants might leave the network or challenge the newly created blocks and its transactions. This protocol brings both advantages and disadvantages if you compare this to the other protocol implementations. The main risk with PoA is that if there is only one validator node, you centralize the risk to a single point of failure. This is a main risk to take into consideration when we talk about distributed networks. However, it does not require the massive computing power that is necessary for networks that make use of proof of work. PoA also has an advantage over proof of stake. With PoA the entire identity of a node is put forward. If it acts maliciously it stands to lose his entire stake into the network. With proof of stake, the participant only stands to lose his current stake that he put forward. Which means that someone who has a lower overall participation in the network stands to lose less than someone who has invested heavily in the network.

1.21   PBFT

Practical Byzantine Fault Tolerance (PBFT) is a consensus protocol that is being used in consortiums where the members in the network can at least be partially trusted. It was introduced already in 1999 by Miguel Castro and Barbara Liskov in their paper that carried the same name. The network relies on a "primary" node and all the other nodes act as backups.[57] They are all in communication with one another in order to achieve consensus. There is a lot of communication in the network as nodes not only communicate all with one another but also want to confirm that the message really came from the node claiming to have sent it but also wants to verify that the message wasn't altered during transmission. This protocol can withstand 1/3 malicious nodes (also see the Byzantine Generals Problem) before becoming faulty so a larger network brings with it more security. The main concern of this protocol is that it isn't seen as very scalable because of the huge amount of messages that has to be sent out. It is also a consensus algorithm that might be vulnerable for the Sybil attack. Again, this is why this consensus algorithm only seems to work for a small group of participants which trust each other to a certain level. Positive note is that there is a major reduction in the computational cost of this consensus algorithm (compared to for example proof of work).

[57] Curran, B. (April 18, 2020). What is Practical Byzantine Fault Tolerance? Complete beginner's guide. *Blockonomi*. https://blockonomi.com/practical-byzantine-fault-tolerance/. Accessed July 18, 2019.

1.22   THE CUCKOO CYCLE[58]

The cuckoo cycle is a proof of work algorithm that is aimed to be ASIC resistant. This specific algorithm was designed when the cryptocurrency craze (2018) was at a height and a lot of companies and people were investing in ASIC miners and pushing other participants out. Blockchain was always intended to leave space for all participants who have general purpose computers but with all these participants, they were pushed out for those specializing in miners. So there was the rise of resistant protocols such as the cuckoo cycle. It was created by John Tromp and is fit for GPU mining while it focuses on memory use, rather than drawing out the GPU speed. This makes it also an energy-efficient algorithm to use. It is based on a graph theory-based algorithm where it tries to find a fixed length L ring in the cuckoo cycle bipartite graph randomly generated by siphash.[59] As the scale of the graph increases, the L value increases and becomes more difficult to find. There are also two alternative proof of work algorithms based on the Cuckoo cycle called CuckAToo and CuckARoo. The first is designed to be more ASIC friendly while the second is made to be even more ASIC resistant.

1.23   DECOR+HOP

Proposed by Sergio Demian Lerner, DECOR+HOP is a protocol designed to help the blockchain to easily scale while still being Byzantine fault tolerant.[60] We can split this approach in two different parts: DÉCOR+ (Deterministic Conflict Resolution) and HOP (Header Only Propagation). The first is a reward sharing strategy while the second (you might have guessed it) combines several elements such as propagating the headers first but also mining on unverified parents (SPV mining, later explained in the Bitcoin chapter). The main changes that DECOR+ brings to the existing "classic" proof of work consensus protocol is the way it wants to resolve conflicts. In the classic approach, there is a determination of which block is correct in case of conflict and the miner that mined the correct block will consequently receive the reward. With this new approach there is a much wider field being considered. The resolution that is being achieved should maximize the revenue of all miners involved, both for the conflicting miners and the rest. In combination with HOP this can be done with a high speed as it is not block size dependent but it evolves with the logarithm of the network diameter.

[58]   Tromp, J. (November, 2019). Cuck(at)oo cycle. *Github – cuckoo*. https://github.com/tromp/cuckoo. Accessed July 22, 2019.

[59]   Oscar, W. (March 22, 2019). WTF is Cuckoo Cycle PoW algorithm that attract projects like Cortex and Grin? *Hackernoon*.   https://hackernoon.com/wtf-is-cuckoo-cycle-pow-algorithm-that-attract-projects-like-cortex-and-grin-ad1ff96effa9. Accessed July 25, 2019.

[60]   Lerner, S.D. (November, 2014). DECOR+HOP: A scalable blockchain protocol. *Semantic Scholar*. https://pdfs. semanticscholar.org/141e/d5f15e791ec7a9537a7b3250f4b7524ce302.pdf. Accessed July 27, 2019.

1.24   GHOST—SPECTRE—PHANTOM

An important protocol that we will often see in "modern" implementations is the GHOST-protocol or "Greedy Heaviest Observed Subtree" "The concept was introduced in 2013 by Zohar and Sompolinsky and put forward in their paper "Secure High-Rate Transaction Processing in Bitcoin" as an answer to some of the problems that are created when a blockchain platform makes use of a higher block time. It combats both the high number of stale blocks that are propagated through the network and the centralization bias. How does it achieve this? By actually adopting the stale or "orphaned" blocks as "uncle" blocks. This way the miners that mined such a block still receive a partial reward and selfish mining can be prevented. The uncle blocks can be included in the main chain, increasing its security. Therefore, the longest chain is no longer the most important one, it is the chain that has the most computational effort that becomes the main chain. You can see that here we still have the added value of faster block time while it does not affect the security of the blockchain itself. In case of conflict, the main chain thus becomes the one with the heaviest subtree rooted at the fork.[61] In its core it is still a proof of work consensus protocol that wishes to increase the security when you want to increase the scalability and speed of the network.

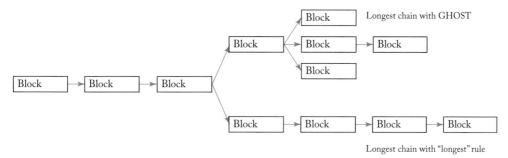

Figure 1.15: GHOST.

   When we start to move to BlockDAG structures (explained below), other consensus protocols can be used to achieve consensus in the network. SPECTRE or "Serialization of Proof-of-work Events: Confirming Transactions via Recursive Elections" is proposed by Sompolinsky, Lewenberg, and Zohar.[62] With this implementation, the concepts of mining and consensus are partially split from one another. On the mining level, we still work with a proof of work protocol that makes no assumptions about the transactions that are actually included in the blocks. This way we can create new blocks at a high speed (the network speed). To achieve consensus SPECTRE

[61]   Sompolinsky, Y. and Zohar, A. (August, 2013). Secure high-rate transaction processing in bitcoin. *IACR*. https://eprint.iacr.org/2013/881.pdf. Accessed July 30, 2019.

[62]   Sompolinsky, Y., Lewenberg, Y., and Zohar, A. (2016). SPECTRE: Serialization of proof-of-work events: Confirming transactions via recursive elections. *HUJI*. www.cs.huji.ac.il/~yoni_sompo/pubs/17/SPECTRE.pdf. Accessed August 1, 2019.

looks at the transaction level completely separate from the mining of the blocks. There is a recursive voting system in place based on the precedence of the blocks where every block submits a vote for every pair of blocks. While looking at the pair, every block votes either -1, 0, or 1 depending on which block they prefer. In example, if there are two blocks A and B, and block C has to vote. It will vote 1 if C prefers A, vote -1 if C prefers B, and simply 0 if there is no preference. This creates an additional ordering on top of the topological ordering of the blockDAG structure and gives us an idea of which transactions can be seen as confirmed and verified within the network.[63] This voting occurs based on the vision that block C has of the network. If block C only exists in the future of block A (it has been built on top of A), of course it will vote for A. If C exists both in the future of A and B, it will look at the past and determine which block has the most support in the network. And even if C doesn't exist in the future of either A and B, it will cast a vote according to the entire blockDAG structure. One of the most important aspects that one needs to consider is time. Transactions are bound to a certain timing and attackers often try to hide a certain block that they propagate later through the network. With SPECTRE this leads to no profit whatsoever because in case of conflicting transactions, the oldest one receives precedence. When blocks become visible, they connect to other blocks within the blockDAG in such a way to other honest block will quickly be able to determine which block has the truthful transactions. With SPECTRE we get also the introduction of the "weak-liveness" concept because every non-faulty node their transactions are being accepted (assuming no conflicts). So we have a nonlinear network of blocks in the DAG-structure. A limitation of the SPECTRE protocol is that it can only be used with cryptocurrencies or networks where a strict ordering of the transactions is not a necessity (we refer to Condorcet's paradox explained in the beginning). PHANTOM is very similar to SPECTRE but it assumes a strict ordering of blocks and transactions in the overall system. This comes with both advantages and disadvantages. We lose here the risk of Condorcet's paradox and we achieve a system that can be used for smart contracts but on the other hand we have to give in on the speed that can be achieved by a consensus protocol such as SPECTRE. So, while mining remains similar to the SPECTRE-system in place, PHANTOM will take another approach when it comes to consensus because it will look for a correct blockchain inside the entire blockDAG. Within the blockchain the ordering of the transactions is enforced by the recursive algorithm being used.[64]

[63] Stone, D. (March 26, 2018). An overview of SPECTRE—a blockDAG consensus protocol (part 2). *Medium*. https://medium.com/@drstone/an-overview-of-spectre-a-blockdag-consensus-protocol-part-2-36d3d2bd-33fc. Accessed August 3, 2019.

[64] Sompolinsky, Y., Wyborski, S., and Zohar, A. (February 2, 2020). PHANTOM and GHOSTDAG. A scalable generalization of Nakamoto Consensus. *IACR*. https://eprint.iacr.org/2018/104.pdf. Accessed February 27, 2020.

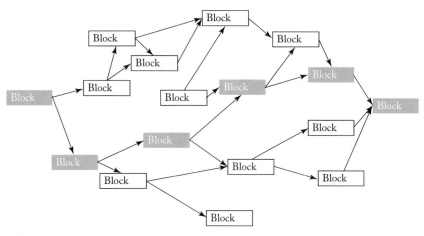

Figure 1.16: Blockchain in the blockDAG.

Attackers are diverted in two important ways. Again, withholding a block from the network is not beneficial as the network continuously keeps on building interrelated blocks and a block that has been kept from the entire network for too long simply can no longer wreak real damage because the "honest" blockchain will clearly have the longest structure. The attacker could also try to build the longest chain by mining faster than the other miners in the network but to achieve this, the attacker should have at least 50% of all computational power in the network.[65]

1.25   ETHASH ALGORITHM AND DAGGER HASHIMOTO[66]

Dagger Hashimoto was the first research implementation of the proof of work mining algorithm used in Ethereum 1.0. The goal of this protocol was double: on the one hand it aimed to be ASIC-resistant so that the benefit of using specialized hardware was reduced to a minimum. Ethereum aimed to be a network that was accessible for all users, also those that couldn't afford expensive hardware. On the other hand, a light client should have been able to verify the mined blocks with relative ease.[67] It was based on two previous algorithms: Hashimoto and Dagger. The first was developed by Thaddeus Dryja and its goal was ASIC-resistance. Dagger was the algorithm that wanted to achieve memory-hard computation but memory-easy validation. Several approaches

[65]   Stone, D. (March 29, 2018). An overview of PHANTOM: A blockDAG consensus protocol (part 3). *Medium.*   https://medium.com/@drstone/an-overview-of-phantom-a-blockdag-consensus-protocol-part-3-f28fa5d76ef7. Accessed August 4, 2019.

[66]   Ray, J. (April 2, 2019). Welcome to the Ethereum Wiki! *Github – Ethereum.* https://github.com/ethereum/wiki/wiki/Ethash and https://github.com/ethereum/wiki/wiki/Dagger-Hashimoto. Accessed Augst 6, 2019.

[67]   A third goal was that miners should store a full copy of the blockchain but for this, some modification tot he algorithm were necessary.

were tried such as "Blockchain-based proof of work"[68] and "random circuit".[69] The advantage of adding Dagger Hashimoto over Hashimoto is that a custom-generated 1 GB data set is used as data source instead of the blockchain itself. This data source is updated based on block data every N blocks by making use of the Dagger algorithm. Eventually, it would be overtaken by Ethash which originated from the Dagger Hashimoto protocol but was changed drastically. This protocol starts with a seed that can be computed for each block by scanning the block headers, from this seed a 16-MB pseudorandom cache can be computed. From this cache, a 1-GB dataset is generated with each item in the dataset depending on only a small number of items in the dataset. When you are mining, you grab random slices of the dataset and hash them together, where verification can be done by using the cache to generate the pieces of the dataset you need. The large dataset is updated once every 30,000 blocks.

1.26   KECCAK256 / SHA3[70]

Keccak-256 was the hashing algorithm that won the competition of NIST to become the SHA3 algorithm. There is one important difference, though: NIST changed the padding, so that SHA3-256 and Keccak-256 have different outcomes even though the underlying algorithm is the same. To this day, Keccak is being used in the Ethash proof of work protocol in the Ethereum network but it is also used by other networks such as Monero where it is not used as a proof of work protocol but rather helps with random number generation, block hashing, transaction hashing, and much more. Keccak is based on the sponge construction, which is a mode of operation where a fixed-length permutation and a padding rule is used so that a variable-length input can be translated to a variable-length output. This translation consists out of XORing of the message blocks that are being used as input, into a subset of the state which is then transformed as a whole by making use of a permutation function f (this is also called the "absorbing" phase). The next phase, called the "squeeze" phase, output blocks are read from the subset of the state and alternated with the transformation function f.[71]

---

[68] Based on running contracts on the blockchain but proved to be vulnerable to long-range attacks (think trap-doors).

[69] Developed by Vlad Zamfir and its goal was to generate a new program every 1,000 nonces, choosing a new hashing function each time, faster than FPGAs can reconfigure. Difficulty is to generate random programs general enough so that there are no gains from specialized hardware.

[70] Bertoni, G., Joan, D., Hoffert, S. Peeters, M., Van Assche, G., and Van Keer, R. Keccak specifications summary. https://keccak.team/keccak_specs_summary.html. Accessed August 7, 2019.

[71] The size of the part of the state that is written and read is called the "rate," while the size of the part that remains untouched is called the "capacity." It is this capacity that determines the security of the algorithm.

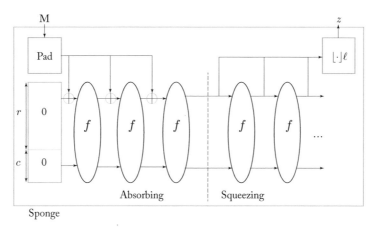

Figure 1.17: The Sponge Construction.

1.27   OTHER PROTOCOLS USED IN BLOCKCHAIN PLATFORMS

Several other types of protocols can be used in blockchain platforms so that one can come to a consensus. One of these protocols, strongly related to proof of stake, is proof of importance. The difference with proof of stake is that in the PoI environment, also the transactions of the user are taken into account. This way the protocol tries to measure the level of trust and importance of the node in the entire network.[72] Another interesting protocol is the Proof of Activity protocol, related to both proof of work and proof of stake. It's more energy efficient than proof of work as only in the first phase this is used, as in the second the protocol makes use of proof of stake. There is also proof of capacity, where the main driver is the hard disk space that is still available (instead of CPU as we find with the proof of work protocol).[73] Other protocols that you might encounter are: proof of replication, proof of burn, proof of space, proof of space-time, proof of deposit, proof of data possession, and so on. You can clearly see that that a lot of different blockchain platforms are experimenting with different solutions to provide consensus in a distributed and decentralized environment, in a secure and efficient way. For each of these protocols you can give both advantages and disadvantages, depending on the goal you are trying to achieve and the way you are working with your organization. In the following chapters we will meet several of these protocols and others and we will go into more detail on how these work.

---

[72]   Used by the NEM blockchain.
[73]   It is also called hard drive mining and can be found, i.e., with the Burstcoin cryptocurrency.

1.28   NONCE

The nonce is a term that finds its source in cryptography. It is an arbitrary number that can only be used once to be used in cryptographic communication. The source of this number is more often than not a (pseudo-)random number generator and is used within communication to prevent for example replay attacks.[74] When we speak about blockchain, we often speak about the nonce. In Bitcoin and its proof of work algorithm where it plays an important role for the miners involved in the network. While the serialized block headers in Bitcoin are in an 80-byte format, the nonce is a 4-byte field. The number in the nonce can be modified as needed so that the header hash is lower than or equal to the value set by the network difficulty. When there is a solution found that is acceptable, we say that the "golden nonce" has been found. In practice, this often means that mining applications will look for a nonce that results in a block hash with 32 leading zeroes. Important to know is that the nonce shifts the workload to the searching of the correct hashing value and makes it much easier to verify a found hash. Because the output of a hashing function cannot be easily predicted based on the input, mining involves a lot of trial and error until an acceptable hash can be found. The nonce used within the Bitcoin network has changed over time. In the beginning, a miner could iterate through the nonce until he found the correct solution. With the increasing difficulty, it happened that miners went through all values of the nonce without finding a solution so that they had to update the timestamp in the block header to account for the elapsed time, which again let to different results among the miners. With the increasing computer power, also this approach became difficult as the nonce values were exhausted in less than a second.[75] A new source of chance was thus necessary to make sure that mining could continue within the Bitcoin network. The solution that was brought forward focused on the coinbase transaction and to use this space as a source of extra nonce values. This allows the miners to explore 8 bytes of extra nonce on top of the 4 bytes of standard nonce. If in the future the miners are able to completely cover this space as well, we could once again work with adjusted time stamps and even use the coinbase script to allow for more nonce space. Now one important question remains: how is this nonce propagated through the network? The answer is: by making use of a gossip protocol. A gossip protocol works like a virus would in an epidemic, or how gossip spreads. Person A tells person B a piece of gossip, both A and B tell it to C and D, and so on. You can see how gossip, or in this case the nonce, spreads through the network. It is used by Bitcoin, Hyperledger implementations, and Hashgraph to spread information through the network (not always the nonce!).

---

[74] A replay attack is an attack where the attacker has found his way within the network. There he waits for a data transmission and tries to delay or repeat it. If there is no way of knowing that a communication already took place, the attacker is successful. A nonce is a possible way to help prevent this attack, as it can be used only once.

[75] The hardware mining computer power started to exceed GH/s and with ASIC mining, we even entered the TH/s hash rate.

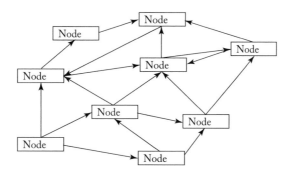

Figure 1.18: Gossip protocol at work.

1.29   BLOCKCHAIN FORKS

Blockchain forks are an important subject within the world of blockchain. It refers to competing or coexisting side chains within the same network. Simply because of the decentralized structure of the network, the occurrence of forks seems to be natural. Blocks are propagated through the network and arrive at different nodes at different times. This is can also be the cause of the so-called orphan blocks. Normally, the nodes will try to extent the chain with the largest cumulative difficulty.[76] We can talk about a fork when there are two or more candidate blocks that are competing with one another to form the longest chain. If a miner discovers a "correct" block, it is immediately sent to its neighbors. Several nodes can in time discover a different solution and broadcast this through the network. The nodes closest to the original miners of the block will start building their chain based on this block and continue working on next blocks. If a fork comes into existence this way, the issue is normally resolved within one block. The reason is that one group of miners will find a next solution first, even if the computer power within the network is evenly distributed among the several competing groups. The next solution will be shared among the network nodes, accepted and spread through the network. The competing nodes will receive this next solution, accept it and stop working on the competing solution, thereby resolving the fork. [77]

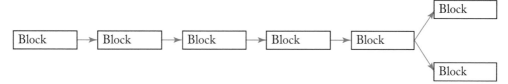

Figure 1.19: Blockchain fork.

---

[76]   The chain that contains the most proof of work.
[77]   A fork like this might happen once a week while a fork that extends to two blocks is extremely rare (because of the explanation above).

There is also the occurrence of hard forks. This is when there is a software update over the network where protocols or mining procedures are upgraded. Once the upgrade has happened, transactions that are being mined by making use of the older software, will no longer be accepted by the upgraded nodes. This way a new and persistent branch comes into being. There comes a parallel set of transactions into being that take place on the different chains. A soft fork is a change in the software where only previous blocks and transactions are made invalid while still being backward-compatible going forward. Another difference between a hard and a soft fork is that for a soft fork only a majority of the miners need to upgrade while a hard fork requires all nodes to upgrade to the new version.

1.30   SIDECHAINS

With the explanation of forks, you can start to imagine the existence of sidechains. These are blockchains attached to another (the "parent" and the "child") by making use of a two-way peg. Because of this connection, assets are interchangeable over the network at a fixed deterministic exchange rate while the sidechain can operate completely independently of the parent and make use of its own consensus protocol.

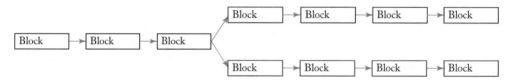

Figure 1.20: Sidechains.

This transfer is in fact nothing more than an illusion. Tokens are locked in the parent chain and the equivalent amount of tokens are unlocked in the child chain. If you want to transfer, the tokens in the child are locked while the cryptocoins in the parent are once again unlocked. For this to be possible, there are several assumptions that are made. The most important underlying principle that we want to reach and understand is the point of something called "settlement finality" (which we explain in more detail in Chapter 2). Its practical implications mean that we have to trust in the honesty of the participants in both chains and that they are both censorship resistant. All of this requires that the participants are honest, including those participants holding the locked tokens. Otherwise you enter a situation where locked tokens can be spent and we create a situation where double spending is once again possible. There also exists the possibility where the child chain doesn't have settlement finality. In this case one could make use of so-called custodians that have to vote when to lock or unlock a certain amount of tokens. This voting system can be adapted to any form which suite,s the blockchains that are being linked the best which makes this quite a flexible system to work with. There are several ways that this system can be implemented. The first is by making use of a central exchange that enforces the two-way peg between the two chains by only unlocking coins of chain 1 when an equivalent amount of tokens belonging to chain 2 are locked.

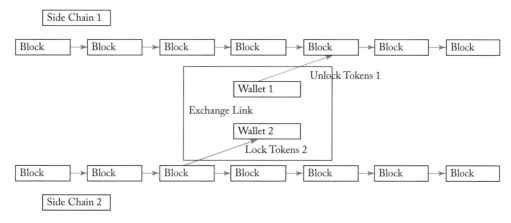

Figure 1.21: Central exchange.

You can clearly see that using this system goes against the very nature of blockchain. This way you are reintroducing the single point of failure and you are once again making use of centralization. You could try to set up a form of decentralization by making use of multiple parties that make use of a multi-signature approach. This is something that could perfectly work in a private setting.

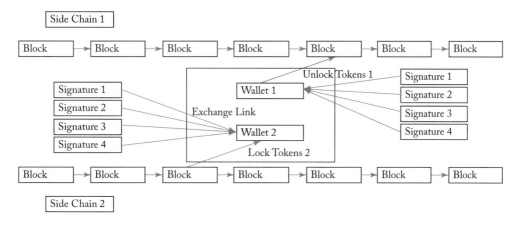

Figure 1.22: Multi-signature approach.

A second approach is stepping away from any centralization and linking the two chains by implementing an understanding of each chains consensus system. This way the tokens can be unlocked from the second the chain is able to verify that there has been a locking transaction. This brings several insecurities with it when you are working with a system where one of the chains doesn't have settlement finality. This is something that again could be applied in a private blockchain/distributed ledger setting, but not in the public world considering the risks that this set-up would bring in what is essentially a trustless environment. You could make use of several ways to create this specific set-up, but it would have to come down to a simplified way of acknowledging transactions and therefore make use of the Merkle root that is so often used in one way or another in the blockchain world.

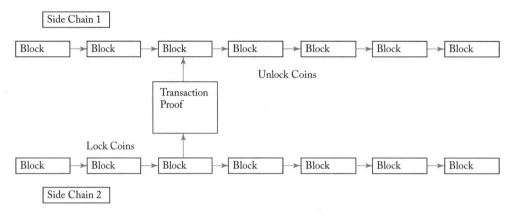

Figure 1.23: **Linked by consensus.**

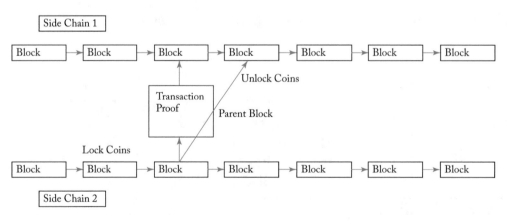

Figure 1.24: Entangled chains.

Another approach that is related to the previous example is called "entangled blockchains." Here the relationship between the two separate chains is brought to the next level. When coins are locked in one chain, this immediately means that the equivalent amount in the other chain are released and vice versa. There are several ways to achieve this. The easiest might be to lock this in metadata within the transactions themselves (this is what we will see later on in the link between Counterparty and Bitcoin as the OP_RETURN opcode is used to lock in certain information).

Other ways exist out of using multiple parents for each block in the second chain or anchoring by cryptographic means in the transactions.[78]

The final example we are going to give is that of "drivechains." Here the participants are allowed to vote on when to release the locked coins and when to send these to another chain. These votes can be locked within a certain section of the transaction information. These voters are more often than not linked to one of the chains, determining the actions that take place for the other chain as well. You can clearly see that trust in the participants is the main concern here.

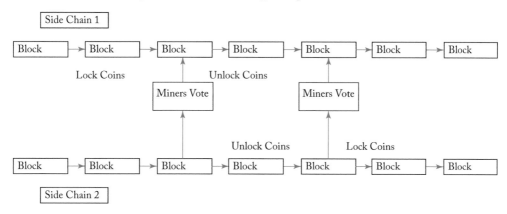

Figure 1.25: Drivechain.

Of course these are all just clear-cut examples that can be used independently but in reality a number of combinations between these approaches is a real possibility. Depending on the use case you are working on, splitting or combining these approaches can fit your solution best. A lot depends on the private–public and permissioned–permissionless approach, combined with the expected trust in the participants.

1.31   BLOCKCHAIN EXECUTION ENGINE

First, before we go into the explanation of the blockchain execution engine, you should understand two concepts: the state machine and the virtual machine. In short, a state machine records every state and will process changes to this state. Explained with a simple example: if you have 10 euro in your wallet, this is the "starting" state. You decide to spend 2 euro, there is a new state. To be able to process the transactions taking place, a blockchain makes use of a virtual machine which is capable of executing the instructions which are encoded within these transactions. The Bitcoin transaction validation engine is quite simple as it relies on two types of scripts: a locking script and an unlock-

[78]   (2015). Sidechains, drivechains, and RSK 2-Way peg design. *Rootstock*. https://www.rsk.co/noticia/sidechains-drivechains-and-rsk-2-way-design/. Accessed August 12, 2019.

ing script). This locking script (or scriptPubKey) specifies the conditions that must be met so that the output can be spend in the future. The unlocking script (or scriptSig) satisfies the conditions specified by the locking script so that the output can be spend. The Ethereum virtual machine on the other hand aims to be a runtime for executing general-purpose smart contracts. There are currently 140 unique opcodes that can be used to execute a specific set of instructions. With the Serenity update, a new EVM (Ethereum Virtual Machine) will be implemented which is based on WebAssembly.[79] Currently, the perception is that this will be EWASM (Ethereum-flavored Web-Assembly) because it leverages improved hardware features and can be built on a wide ecosystem of tooling and language support. It is "Ethereum-flavored" because it has to be deterministic and includes several smart contracts that provide access to specific Ethereum platform features.

1.32   SERIALIZATION

A concept that you will get familiar with later on in the book is "serialization." In the world of computer science and networking, serialization helps you a lot to get ahead. It basically refers to how you are going to store certain data structures and how you are going to transmit these over the network. Certainly in distributed networks this is of importance (refer back to the concept of block time). You need an efficient and secure way to transmit the data. Over time this might change and new formats can be used to help improve security, accuracy, or other implementations in the network. Sometimes these new formats can be backward compatible, at other times the new format is forces through the network (and here we have hard and soft forks again). Of main importance to remember for now is that there is a structured way in each network to store and transmit the data. These can differ significantly between the different platforms based on the choices that were made by the developers.

1.33   THE BLOCKCHAIN TECHNOLOGY STACK

We are going to talk about the blockchain technology stack in way more detail as we progress in this book but it might be good that you already have a general idea of how new applications are being built when we talk about blockchain and how the architecture (in a very general sense) might look like. Opposite to other technology implementations, you have to consider the entire "stack" of the technology when you would like to work with it within your organization. At the core there is decentralization and consensus that you would like to consider and so you have to look at your very infrastructure and ask yourself the question: are you currently prepared to step into a new way of working? You have to, in a sense, let the classic view of centralization and control go to make room for an interconnected system that no longer has a single point of failure.

---

[79]   WebAssembly is an open standard which provides an optimized binary format which is supported by several runtime environments so that it is executable in most modern web browsers.

Layer	Description	Examples
Application	User interface	dAPP, user interface, chaincode, etc.
Services	Interconnection of applications	Oracles, wallets, smart contracts, etc.
Protocol	Consensus protocol	Algorithms and side chains
Network	Transportation of information	P2P, PRLx, etc.
Infrastructure	Node infrastructure	Mining, tokens, nodes, storage, etc.

It comes with both advantages and challenges but each of these "layers" has to be taken into consideration when you are thinking about applying blockchain. It is much more than cryptocurrencies alone, as you will soon discover. So keep this image somewhere in the back of your head as we are starting our journey.

1.34   DAG—DIRECTED ACYCLIC GRAPH

I was in doubt when I was writing this chapter if I should already add a chapter on DAG or not. I decided to do it because, even though it might seem a little confusing, soon you will clearly understand what this technology can do and a discussion about DAG never seems to be far away in the current IT landscape. So, what are we talking about? DAGs are another form of distributed ledger technology, just as blockchain is. The main difference lies in the fact that in the world of DAG there are no more blocks. This might seem confusing as I just try to tell you that everything in the world of blockchains is basically "chained blocks" but, of course, as with everything in life, things just aren't that simple. With DAGs the transactions are directly linked to each other. Not in a neat little row, more in a cloud of transactions that link to a couple of new transactions, and so on. It actually just tells you everything that you need to know in the name of the technology itself. It is directed, which means that (as you can see in the image below) all links point in the same direction and because of this there will be no loops possible within the network. It is therefore "acyclic."

It basically tries to offer the same functionalities of blockchain but with a better performance.[80] It offers on top of that better scalability and lower transaction fees (because there are no miners in the network). Contrary to what you have learned from blockchain technology, here the network will start to work faster as the number of transactions that wait to be validated starts to increase (minimizing the possibility of congestion of the network). The main concern that currently people are working on is the best manner to reach a secure decentralized consensus within the network. When the answer is found on this question, DAGs might pose a "threat" to the current blockchain landscape because of the advantages they bring. It is also often called "blockchain 3.0" as it is seen as the next natural step in the world of decentralized applications and the future way

[80]   Thake, M. (November 9, 2018) What is DAG distributed ledger technology? *Medium*. https://medium.com/nakamo-to/what-is-dag-distributed-ledger-technology-8b182a858e19. Accessed August 14, 2019.

of working. As always, this was just a very broad base on what DAGs are, while in reality, each and every implementation differs greatly and comes with its own advantages and disadvantages. Later on in the section, we will look in more detail in what is currently already there and what will be possible in the near future.

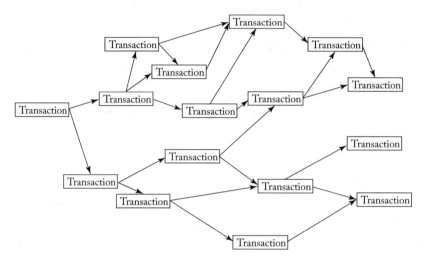

Figure 1.26: DAG at work.

1.34.1 MERKLEDAG

And while we have entered the world of DAGs, why not immediately explain what MerkleDAGs are. This is a Merkle Directed acyclic graph, which has a similar structure as a Merkle tree with that difference that it does not need to be balanced and the non-leaf nodes are also allowed to contain data. The edges are constructed as Merkle-links which means that these links can be used to identify the objects that they are linking to.[81] This way it allows for uniquely defined cryptographic hashes which are tamper proof and makes sure that there is no duplication of data. It transferred, you can send huge amounts of data to another person.

1.34.2 BLOCKDAG

We also have the blockDAG paradigm. When you understand how blockchains and DAGs function on a high level, this concept will be quite clear. We still work with blocks but there is no longer a single parent. Instead, each block references all tips of the graph that the miner can observe locally. You might immediately see the issue here. As there are many blocks and many branches of blocks

---

[81] Batiz-Benet, J. (2018). go-merkledag. *Github – ipfs*. https://github.com/ipfs/specs/tree/master/merkledag. Accessed August 14, 2019.

referencing to each other, the possibility certainly exists that there are conflicting transactions that are part of the same chain. Therefore, it needs specific consensus protocols to achieve some sense of security over the entire network. There have been some protocols that already have been developed (such as SPECTRE, PHANTOM, and Inclusive).

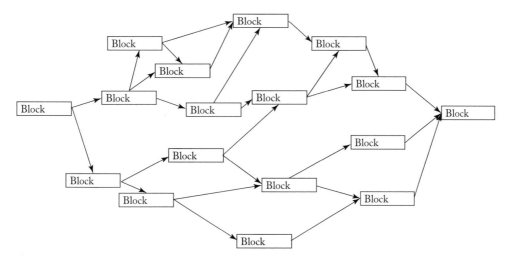

Figure 1.27: BlockDAG structure.

With a functioning blockDAG network, you could speed up transactions to seconds, reduce fees to a minimum, and support decentralization of your network (because a lot more blocks can be mined), there are less orphaned blocks and the incentive for selfish mining is strongly reduced.[82] BlockDAG-based networks try to find a suitable relationship between the networks that were devised by Satoshi Nakamoto, Vitalik Buterin, and others and DAGs. An example of a network that implements this technology is Soteria.

1.35   BLOCKCHAIN-SPECIFIC ATTACKS

While the Bitcoin network and other blockchain implementations (or distributed ledger implementations for the purists among you) try to solve the Byzantine Generals Problem, still other attacks are possible on these types of networks.

The 51% Attack

A famous one is the 51% attack. The principle behind this is simple: if you are able to gain control over at least 51% of the nodes on the network, you could drive the consensus needed to a lie. The network would accept the lie that you initiated but this

[82]   Tran, A. (May 23, 2018). An introduction to the BlockDAG paradigm. *Daglabs*. https://blog.daglabs.com/an-introduction-to-the-blockdag-paradigm-50027f44facb. Accessed August 28, 2019.

would also mean that you need to retain control over at least 51% at all time. From the moment that you lose control over the majority, the network will be able to push out the lie. You can easily see that such an attack in a large network such as Bitcoin or Ethereum is virtually impossible, this is why this type of attack is mainly used on the smaller cryptocurrencies where it is easier to gain a majority in computer power within the network. In practice of course at this time the attacker(s) have already left the network with their profits. It has happened in the past that attackers have used the 51% attack to start double spending coins or completely stop the acceptance of new payments. Several of these attacks have happened in the past and while the damage in terms of monetary value lead to numbers often in millions, the real damage is the loss of trust in the network. It cannot only damage the network but also completely destroy a cryptocurrency. One of the more recent attacks was the attack on Litecoin cash in June 2018. The damage was limited due to the early detection of the attack on the network.[83] It is important to note that one cannot push out old blocks, unless the networks abandons a part of the chain or fork.

Race Attack

This is another type of attack which focusses on anyone accepting payments on a blockchain network. Here speed is the key for the attack taking place[84] as the attacker will first make a payment to one receiver, for example a merchant, and a second to another merchant or even himself. If the first receiver accepts the payment with 0 block confirmations, the possibility exists that the second transaction is being mined and accepted in the next block while the first remains unmined, which leaves the first without payment for the delivered services.

Finney Attack

The Finney attack is quite an ingenious attack whereby the focus also lies on receivers of payments, specifically (once again) on those that accept payment for a good or a service at 0 block confirmations. The attacker himself is a miner that has a mined block but did not yet broadcast it to the other nodes within the network. He includes within this block a false payment between two addresses that he both owns. Next, he sends a payment to a merchant which accepts the payment before it is confirmed and provides the good or service demanded. In the final stage the attacker finally broad-

---

[83] Several other attacks have happened in the past with as famous examples: Krypton, Verge, Bitcoin gold, Monacoin, Zencash, and others.

[84] This type of attack would not occur for example on a Bitcoin network where mining takes about 10 min per block.

casts the block with the false payment and this payment would be accepted before the payment to the merchant ever has a chance, thereby leaving him unpaid.

Vector76 Attack

The Vector76 attack is a combination of the Finney and Race attacks. There are two nodes in this attack: one connected to the exchange node and one connected to several peers in the network. After this, the attacker has a high-value and a low-value transaction, pre-mines, and withholds the high-value transaction block from the exchange service. After a block announcement, he quickly sends the pre-mined block to the exchange service. Part of the network will accept this block, while another part of the network doesn't see this transaction. Once the high-value transaction is confirmed by the exchange service, the attacker sends the low-value transaction to the main network that finally rejects the high-value transaction. This will end in the attacker's account being deposited the high-value amount.

Sybil Attack

Here the attacker fills the network with his nodes. When new nodes connect to the network, the chance exist that they are only connected to the nodes of the attacker. In the next phase, the attacker stops blocks and transactions that are otherwise shared through the network. Instead, he only shares what he wants and fills the blocks with false information, thereby putting the new node on a separate network.

Eclipse Attack

The eclipse attack differs from the Sybil attack in that sense it does not aim to attack the entire network. Instead, it focuses on specific user/users which the attacker wants to isolate and attack. This allows the attacker to make sure that the user no longer has a clear view of the network and the ledger state. This in turn can lead to double-spend attacks or other types of attacks on the victims.

Selfish Mining

A selfish mining attack (or block withholding attack) is an attack where the miners try to earn more cryptocurrency than they would normally be able to.[85] By hiding newly mined blocks from the other miners and the blockchain, they create a separate fork. This way they create a private fork and by timing when other miners get to see the new fork, they will abandon their own work to join this new fork. This process

---

[85] Eyal, I. and Sirer, E.G. Majority is not enough: Bitcoin mining is vulnerable. *Cornell.* https://www.cs.cornell.edu/~ie53/publications/btcProcFC.pdf. Accessed August 20, 2019.

continues until the private chain is greater than the public chain, showing that it becomes more lucrative to join this separate chain. If the majority would join this private chain, this could lead to the collapse of the decentralized nature of the public blockchain as a whole.

Cryptocurrency Mining Malware

Not really an attack on a blockchain network, but still it deserves being mentioned here. Mining malware is a type of malware that uses the resources of computer systems (any device connected to the internet) to generate revenue for criminals. There are examples of browser-based mining malware since 2011 but they are gaining more and more popularity in recent years.

Outsourcing Attacks

Malicious miners could commit to store more data than the amount they can physically store, relying on quickly fetching data from other storage providers.[86] This is a form of cyber criminality specifically linked to blockchain networks or DApps that focus on the storage of data and/or files.

Generation Attacks

Malicious miners could claim to be storing a large amount of data which they are instead efficiently generating on-demand using a small program. If the program is smaller than the purportedly stored data, this inflates the malicious miner's likelihood of winning a block reward in Filecoin, which is proportional to the miner's storage currently in use.

Maginot Line Attack

This name was coined by Tim Swanson and basically means that if an actor has enough hashing power within the network, he would be able to block transactions or rewrite them as they would like.

P + Epsilon Attack

The P + Epsilon attack is a bribery attack.[87] The basic underlying assumption is that the attacker offers a slightly higher reward than the honest participants would nor-

[86]   Protocol labs (July 19, 2017). Filecoin: A decentralized storage network. *Protocol Labs*. https://filecoin.io/filecoin.pdf. Accessed August 28, 2019.

[87]   Buterin, V. (January 28, 2015). The P + epsilon attack. *Ethereum Blog*. https://blog.ethereum.org/2015/01/28/p-epsilon-attack/. Accessed September 2, 2019.

mally obtain for participation in the network. These bribes can be locked in a smart contracts in which the additional reward "epsilon" is available for those proving to have voted for the bribe. However, there is an important catch: if the consensus of the votes is in favor of the attacker, no bribe is being paid out! Only if the briber loses is the bribe actually paid.

Transaction Malleability Attack

Transaction malleability tries to trick the victim to pay a certain transaction twice. Depending on the blockchain network, attackers can try to change transaction IDs or otherwise to trick the victim into thinking that the first transaction has failed. This can lead to the victim trying to send a second transaction, making him pay twice.

DDoS

Even though not a specific attack for blockchain networks, it is the most common type that these networks are dealing with. By bringing down certain mining pools, wallet providers, crypto exchanges, and so on, they can inflict damage on the network.

Time Jacking

Time jacking is an attack where the attacker tries to alter the network time counter on a certain node by making use of fake peers in the network with inaccurate timestamps. If the attacker is able to convince the node to adjust its network time counter, he can force the node to accept a different blockchain.

Single Shard Takeover Attack

The 1% shard attack is an attack that becomes a possibility in a sharded network environment. In this case an attacker is focusing on a single shard of the network and he uses his hashing power to take over this shard. The idea is that an attacker has 1% of the hashing power of the network, and if there are 100 shards, the attacker can take over 100% of a single shard. With a proof of stake implementation, this attack vector is taken away from the possibilities.[88]

---

[88] Dexter, S. (March 11, 2018). 1% Shard attack explained—Ethereum Sharding (Contd…) *Mango Research*. https://www.mangoresearch.co/1-shard-attack-explained-ethereum-sharding-contd/. Accessed September 5, 2019.

1.35.1 VIRTUAL MACHINES

There are many blockchain networks nowadays that make use of virtual machines.[89] Also these aren't free from possible attacks. A common issue is the loss of tokens when they are sent to an address that no longer exists or when a smart contracts runs but cannot complete its execution. Some platforms have started mitigating action against such flaws but not all of these platforms are already proof against such losses. A second issue is the fact of immutability. Once a smart contract is deployed, it can no longer be changed, meaning that an attacker can make use of the discovered vulnerabilities within a smart contract without the developers being able to protect their users or change their code. Another possibility is access control where attackers can try to get access to sensitive functionalities of smart contracts when a bug is discovered.[90] Finally, there are the short-address attacks. If the virtual machine accepts incorrectly padded arguments, an attacker could use this vulnerability to send crafted addresses to potential victims.

1.35.2 SMART CONTRACTS

One of the most famous examples of vulnerabilities in smart contracts is the attack on the DAO in 2016. Commonly, the vulnerabilities are related to the source code of the smart contract language but developers using, Solidity, for example, are also quite capable of building in vulnerabilities in their smart contracts themselves quite easily. This means that users always have to be aware that smart contracts aren't flawless and developers always have to review their code time and again to prevent abuse of their code.

1.35.3 WALLETS

Wallets can fall victim to several types of attacks. There are the classic attacks such as phishing where an attacker attempts to gain access to the wallet by stealing the user information of the victim. This type of attack is a classic attack vector where the attacker tries to gain control over the victim's assets. In the same line, there are dictionary attacks where the attacker attempts to guess the victim's password and break the cryptographic hash of the victim. More specific attacks on the wallet of a victim can be related to the signature algorithm used by the wallet software. If there isn't enough entropy, the same value can be generated multiple times to create a private key, leading to a vulnerability in the signature itself. Similarly, if the key generation has a vulnerability, an attacker can try to gain access to the wallet of the victim. Even cold wallets aren't completely safe. There is, for example, the Nano S Ledger wallet (which is quite popular) where researchers were able to

[89] Bryk, A. (November 1, 2018). Blockchain attack vectors: Vulnerabilities of the most secure technology. *Apriorit*. https://www.apriorit.com/dev-blog/578-blockchain-attack-vectors. Accessed September 7, 2019.

[90] Monahan, T. (2017). Unprotected function. *Github – Crytic*. https://github.com/crytic/not-so-smart-contracts/tree/master/unprotected_function. Accessed September 14, 2019.

perform an Evil Maid attack on the bugs in the software in the wallet and obtain user information based on this attack. [91]

1.35.4 OTHER ATTACKS

This list is certainly not definitive and will change and increase over time. Cyber security is a living subject more than ever and the cyber war has never been waged so massively as in recent years. There are certainly more attacks to come but this doesn't mean that there is no value in blockchain technology; it just means that it is vulnerable to attacks just as any other technology.

[91]   Mihov, D. (February 6, 2018) All ledger wallets have a flaw that lets hackers steal your cryptocurrency. *The Next Web*. https://thenextweb.com/hardfork/2018/02/06/cryptocurrency-wallet-ledget-hardware/. Accessed September 26, 2019.

CHAPTER   2

Bitcoin

2.1   HOW DOES BITCOIN WORK?

From all cryptocurrencies and blockchain networks, Bitcoin was the first and arguably to this day the most famous of them all. It was the first coin that based itself not on a banking system, a centralized government and/or a payment system but is in fact completely decentralized. The network is completely trustless and trust is based on the actions of all the nodes in the network. Consensus is needed to move forward and is the lifeblood of the network. Bitcoin is also the proponent of what is currently called "blockchain 1.0" or the first type of blockchain and blockchain implementation to see the light of day (Unibright.io, 2017). Within the network the proof of work algorithm is being used, more specifically it is based on the secure hash algorithm 256, better known as SHA-256 (see earlier for detailed explanation). The transactions in the network are processed by the miners which generate the hashed output from a block header that is used as input together with a nonce.[92] Why mine for these blocks? It costs time, hardware and energy to find the correct solution of a block. The winning miner receives a number of Bitcoin as reward for his work.[93] This is the high-level story but of course you wouldn't be reading this if you were just happy with the high-level story. We are going to dive into the details and see what makes the Bitcoin network so unique. I would advise all of you to read on if you are interested in development and a technical understanding of blockchain. Even if you just "skim" through the chapter that follows, you will better understand the choices that were made by other platforms and why there are certain evolutions in the world of blockchain. For those of you that just can't get enough of Bitcoin, I gladly refer you to *Programming Bitcoin: Learn how to Program Bitcoin from Scratch* by Jimmy Song (2019) where you get an even more in-depth explanation of how the network works (and how it is programmed), or *Mastering Bitcoin* by Andreas M. Antonopoulos (2017). I chose a different approach than other books on the subject of blockchain as I will work "top-down." I will start with what most people already know: the chain of blocks. I will work my way down from the chain, toward the blocks themselves and eventually the transactions with signing and verification. This means that sometimes you might see words and concepts pop up that might not be immediately clear to you, but you don't need to worry because everything will become clearer as you go on.

---

[92]   The hash value should be lower than the target hash value set by the network.
[93]   At this time a miner is awarded 12.5 BTC per block but this has been decreasing over time.

2.2   THE BITCOIN BLOCKCHAIN: THE NETWORK

The Bitcoin network is an extensive network in which a heap of nodes are constantly communicating with each other. Transactions and blocks are broadcasted throughout the network while the gossip protocol helps to spread the nonce over all the nodes. But first things first: how do we get inside the network? When we have a node there must be a way to discover other nodes in the network and start to make a connection, otherwise it would be futile to start up a node in the first place. To be able to join the extensive Bitcoin network, only one other node needs to be discovered. That's it. When that happens, you are part of the Bitcoin network. The nodes typically connect to each other by making use of a TCP connection over port 8333[94] (or another in case another is provided).[95] New nodes can be discovered by making use of so-called seed nodes which help to quickly discover other nodes in the network. Once our new node has established a connection with several of these "regular" nodes in the network, it will lose the connection with the seed node. When you are making use of the Bitcoin core client, this option is turned on by default (-dnsseed is set to 1). Other ways for a node to discover the address of another node is by making use of a user provided text file on startup, hard coded addresses in the software of the node, stored addresses in a database, and reading these on startup or by DNS request. The node will keep a timestamp for each of the addresses and this timestamp is updated every 20 min when a message is received from a specific node via the Address-CurrentlyConnected in net.cpp (one of the files making up the Bitcoin Core implementation. When our node finally discovers one of the nodes that is already part of the network, it starts by sending out a version message. This version message contains the following information.

Table 2.1: **Version message**

Name	Description
PROTOCOL_VERSION	Version of the P2P protocol
nLocalServices	list of local services
nTime	Current time
addrYou	The IP address of the "old" node
addrMe	The IP address of our node
Subvert	Type of software on the node
BestHeight	Block height of our blockchain

Based on this version-message the "old" node that was already part of the Bitcoin network will respond with a so-called verack-message. With this message the "old" node acknowledges the version message that has been sent by our new node. Now the "old" node can do the same thing and respond with a version message so that it can be acknowledged by our node. This way the nodes

---

[94]   Running a full node. https://bitcoin.org/en/full-node. Accessed July 1, 2020.
[95]   For the testnet the default port is 18333 and for Regtest it is 18444.

become peers of each other. The next step will be for our node to start sending "addr" and "getaddr" messages through the network. With the first message it sends its own address to the newly connected peer so that this peer can send this address further through the network so that more nodes learn about our node and can start to establish connections.

Name	Description	Field Size	Data Type
Time	The actual timestamp	4	Uint32
Services	Bitfield of features to be enabled	8	Uint64_t
IPv6/4	IPv6 address / IPv4 mapped to v6	16	Char[16]
Port	Port number or network byte order	2	Uint16_t

Our node can ask for a list of addresses by making use of the "getaddr" message the same way. The response on the getaddr is a list of the addresses of which the timestamp isn't older than 3 hr old and there is a maximum of 2,500 addresses. If there are more than 2,500 addresses, there is a random selection amongst the addresses that have a timestamp that is younger than 3 hr. The Bitcoin network is a distributed network and nodes can come and go as they please, therefore our node will have to perform network discovery from time to time. This is not very efficient and would lead the network to send out and receive way too much messages. The problem at hand is solved in two different ways. First of all, our node will only connect to a handful of peers. Too many connections are simply not interesting. The second is even more ingenious. When we would disconnect our node, it will make use of a technique called "bootstrapping" to quickly rediscover nodes and reconnect to the network. How does it do this? By keeping a list of the earlier connections the node had in the network. This way it can easily try to reestablish a connection with a node it knows from before. In case no node can be discovered, it is the seed node to the rescue.[96]

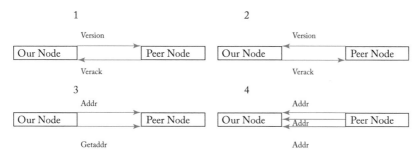

Figure 2.1: The several steps of node discovery.

Of course one can take the more adventurous route and simply turn off the default settings. This is not advisable as the network connections will no longer be automatically maintained. Some

[96] (December 19, 2017) Satoshi Client Node Discovery. https://en.bitcoin.it/wiki/Satoshi_Client_Node_Discovery. Accessed July 1, 2020.

final remarks regarding the nodes and addresses in the network need to be made. When a node receives an "addr"-message, these addresses aren't just simply added. They are checked. It the message is from a node with a really old version, the message is ignored, just as when the version of the other node is not very old (but still older) and you already have 1,000 addresses. The addresses also receive a timestamp in the addr-message and if the timestamp is too low or too high, it is simply set to 5 days ago. Otherwise, we always subtract 2 hr from the timestamp and add the address. The addAddress function will then check if the address already exists and, if so, update the address record accordingly. Now, our node might want to respond to a "getAddr"-message itself. This can be done and "addr"-messages can be sent out but the addresses that are added to the message have to adhere to a certain set of rules:

- after processing, the timestamp of the address < 60 min;

- the address must be routable;

- the addr-message can contain a maximum of 1,000 addresses; and

- fGetAddr is not set on the node.

The node will also broadcast its own address to all connected nodes every 24 hr. The own list of addresses are also being updated and old addresses are removed. This happens every 10 min as long as there are more than 3 connections.[97] A special type of messages that used to be implemented (it has been removed in the more recent versions of Bitcoin core) in most nodes was the "alert message."[98] This was an option which allowed the Bitcoin developers to broadcast an emergency message throughout the network to all nodes. The alert message was used to warn the users if they have to undertake certain actions. Such an action could be an update because of a critical bug in the software in use. So far we have seen discovery messages and alert messages but the bulk of the messages in the Bitcoin network are more "standard" messages. A network message in the Bitcoin network has the following container structure.

Table 2.2: Bitcoin network message

Name	Description	Size
Start string	Start indicator and identifier	4 bytes
Command name	What is the payload	12 bytes
Length of the payload	Length in little-Endian	4 bytes
Checksum field	identifier	4 bytes
Payload	The actual data	Varies

---

[97] (December 19, 2017) Satoshi client node discovery. *Github – Bitcoin*. https://en.bitcoin.it/wiki/Satoshi_Client_Node_Discovery. Accessed November 4, 2019.

[98] Removed with BIP133.

So what information is hidden in the messages? When we look at Table 2.2, we need to understand that the first items are all in the "start string." The network message gives us an indicator of the start of the message that is being transmitted. It also gives us an identifier as each network that makes use of "start string" has its own unique starting code for each message, like a signature. With the use of network magic the messages send out by other nodes from a different network will never be accepted as they do not have the correct "fingerprint." For the Bitcoin main network, this fingerprint is 0xf9beb4d9.[99] The next part is the command field. This gives us a description of what the payload actually is and is meant to be human readable. By examining the command field, we can get therefore a better understanding of what the intent of the message is. The length of the payload is encoded in little-endian and is important for the client to be able to accept a communication or not at all. This can be important as the payload size is variable and can be in some cases too big for a client to accept. The last part of the message header is the checksum field. This checksum field consists out of the first 4 bytes of the SHA256 of the payload. Finally, there is the payload itself. We are not connected to the network. Interesting. What should we do next? You might have guessed it: information about the blockchain, more specifically the block headers. Why the block headers? Well, it is much smaller in size than the actual blocks themselves and already contain the information necessary for a full node to asynchronously download the full blocks. For a light client the block headers are even enough to get proofs of inclusion. We will go deeper into these different node types at the end of the chapter on Bitcoin.

2.3   BITCOIN BLOCKS

To get these block headers, the node sends out a "getheaders" message and this message requests a "headers" message that provides blocks starting from a certain point in the Bitcoin blockchain. It can, however, also be done by making use of a "getBlocks" message. How is it determined which headers the node needs to download? Well, there are two situations: either you have a completely new client which contains no information yet or you have a client that has been disconnected for a while. In case it has been disconnected, the possibility exists that you have information on stale blocks that are no longer part of the "main" chain. The reply to both "getHeaders" and "getBlocks" is a set of header hashes on several heights within the blockchain. Your node will check which one it recognizes from its own chain and start from that point. If there are stale blocks being replied by the peer node, it is up to our node to make the distinction between what is part of the main chain and what isn't. So what does such a reply usually look like? It has the format that looks like this:

---

[99]   For the testnet the start string is 0x0b110907 and for Regtest it is 0xfabfb5da. P2P Network. https://developer. bitcoin.org/reference/p2p_networking.html. *Bitcoin Developer*. Accessed July 1, 2020

Table 2.3: Inventory message that follows getBlocks message

Name	Description	Size
Version	Protocol version	4 bytes
Hash count	Number of header hashes	Varies
Block header hashes	The actual header hashes	Varies
Stop hash	All zeroes	32 bytes

Table 2.4: Inventory message that follows getHeaders message

Name	Description	Size
Count	Number of headers	Varies
Headers	Block headers	Varies

The main difference between getHeaders and getBlocks is the number of header hashes. GetBlocks will get you a maximum of 500 header hashes while the getHeaders goes up to a maximum of 2,000 header hashes.

The block header itself consists of six parameters.

- **Version:** the version number of the Bitcoin software.

- **Parent block hash:** the hash of the last block accepted on the blockchain.

- **Merkle tree root hash:** a hash of all the transactions included in the block.

- **Timestamp:** the time the block was created.

- **Nonce:** a set variable used in proof of work.

- **nBits:** the threshold set by the network.

For the genesis block, we could fill out the following information.[100]

Table 2.5: Simplified block header

Version	Parent Hash	Merkle Hash	Time	Nonce	Nbits
1	0	Genesis.buildMerkleTree()	1231006505	2083236893	0x1d00ffff

So how is this data transmitted over the Bitcoin network? It makes use of the following format to provide some kind of inventory:

---

[100]   The encoding of the genesis block can be found on: https://github.com/bitcoin/bitcoin/blob/3955c3940eff-83518c186facfec6f50545b5aab5/src/chainparams.cpp#L123.

Table 2.6: Data container

Name	Description	Size
Type identifier	Type of the hashed object	4
Hash	Twice SHA256 in internal byte order	32

The data identifier is, as you might already know, the indicator of what data is being transmitted. An example is MSG_TX which indicates that the hash is a transaction identifier. This is the reply to a getData message (or getBlocks/getHeaders message). Here we took the information that is used in the Bitcoin network; as you will see later, this varies depending on the technology implementation. For now, we will use this as a start to gain a deeper understanding of blockchain. Of course the information that has been provided within the block headers has been evolving over time. The genesis-block is an example of the first version that started out in 2009. In September 2012, the block height was the next parameter that was required and it was in March 2013 that blocks that did not contain this information started to be rejected. Version 3 of the blocks was another soft fork which required DER serialization (you will see later what this is) of all ECDSA signatures. Version 4 (which was also released in 2015, just as version 3) was a soft fork that supported the OP_CHECKLOCK-TIMEVERIFY. All these upgrades have been supported by the isSuperMajority() mechanism.

Table 2.7: The structure of a block in Bitcoin

Name	Description	Size
Block size	Size of the block (bytes)	4 bytes
Block header	See below	80 bytes
Transaction counter	How many transactions	Varies
Transactions	The transactions	Varies

An important rule that is included within the Bitcoin network for the serialization of blocks is that the maximum size is 1 MB. So even though there are several variable parts when we look at the size, the total sum has a hard limit that needs to be respected.

Table 2.8: The structure of a block header in Bitcoin

Name	Description	Size
Version	Version of the protocol	4 bytes
Previous block hash	Hash of the parent block	32 bytes
Merkle root	Hash of the root of this block	32 bytes
Timestamp	Creation time of the block	4 bytes
Difficulty	Difficulty target	4 bytes
Nonce	Variable number	4 bytes

The genesis block was created by Satoshi Nakamoto and contained within it a hidden message which not only refers to the earliest time of the creation of the block but also at the same time the philosophy behind the creation of Bitcoin: "The Times 03/Jan/2009 Chancellor on brink of second bailout for banks." Every produced block is then being shared with the rest of the network to be checked and approved or discarded based on the participants their results. Your node can receive blocks based on two methods: either your node has send out a request for block data, where you receive a data message with data type MSG_BLOCK or the blocks can be send out as a part of a "block message" because a peer has just mined a new block. When you are making use of a node, or at least a full client, all the blocks, starting from the genesis block, are being contained within this node. When a node receives a new block, it will start to check if the block follows the criteria: [101]

1. Is the data structure of the block syntactically valid?;

2. Is the block header hash < the target difficulty?;

3. timestamp < 2 hours in the future (for time errors);

4. block size within acceptable limits;

5. first transaction is a generation transaction; and

6. all transactions within the block are valid.

The blocks within the blockchain are linked to each other based on the hash of the previous block. A final part that is crucial to understand is the concept of network difficulty. The difficulty of finding a correct hash can be set by increasing or decreasing the target hash value within the network. This way one can change the rate at which correct hashes can be discovered, the pace at which transactions can be approved and the speed at which new blocks can be added to the blockchain. The average speed of the Bitcoin network has been about 10 min per block. Other networks are much quicker but as we have seen in earlier sections, faster networks come with risks of their own, just as slower networks might have their own vulnerabilities. Every 2,016 blocks the network difficulty can be adjusted based on the length of time it was necessary to solve and find the previous 2,016 blocks. This is done because the network wishes to remain at a pace of 1 block every 10 min but computer power increases over time, so the difficulty needs to be adjusted so that we can maintain the pace of 10 min/block. This recalculation does not happen through a central authority, as this would entirely defeat the purpose of a decentralized and distributed network. Instead, every full node in the network does this completely independently. It does this based on the following equation:

$$\text{Difficulty}_t = \text{Difficulty}_{t-1} * (t \text{ of last 2016 blocks}/20160 \text{ min}).$$

To make sure that the recalculation is not too volatile, the adjustment must be less than a factor 4.

---

[101]   You can find these requirements in the CheckBlock and CheckBlockHeader functions of the Bitcoin Core client.

2.4   BITCOIN TRANSACTIONS

We already have a good idea of how blocks and block headers are being shared and transmitted throughout the network and how our node can ask for this information from its peers. Problem is that we still don't have an idea how the transactions are being shared among the nodes. Here the "mempool" message is being used. With this message our node can request transactions that it has already verified as valid while they have not yet been mined in a block. Here the reply also is an inventory (inv) message which contains the transaction IDs (or TXIDs). This list of unconfirmed transactions might not be complete. If you are running a SPV client (see later), you might receive only those transactions linked to your wallet, nothing else. A regular inventory message can contain up to 50.000 of these unconfirmed transactions, so it can take several messages before your node has the complete list and there are also some problems with the "filterload" message (check out the Bitcoin developer manual for more information as there are a lot more messages that can be transmitted throughout the network). Transactions are being shared among the nodes of the network before they are put inside of a candidate block. Such a transaction has the following structure:

Table 2.9: Standard transaction		
Field	Description	Size
Transaction hash	Pointer to the transactions spent UTXO	32 bytes
Output index	Index of the spent UTXO	4 bytes
Unlocking-script size	Length in bytes	Varies
Unlocking-script	Script that fulfills UTXO locking-script	Varies
Sequence number	Tx-replacement feature	4 bytes

Every transaction that is received by a node must first be verified by that node to make sure that only valid information is communicated to other nodes in the network while the invalid transactions are immediately discarded. This validation of a transaction occurs in a specific manner as there is a specific set of criteria:

1. the syntax and data structure must be correct;

2. $100 \geq$ Transaction size $<$ MAX_BLOCK_SIZE;

3. nLockTime $\leq$ INT_MAX;

4. input and output lists are not empty;

5. inputs should not have hash = 0 or N = -1;

6. for each input, if the referenced output exists in any other transaction in the pool, reject;

7. for each input, check the main branch/transaction pool to find referenced output transaction. If the output is missing, this is an orphan transaction;

8. for each input, if the referenced output transaction is a coinbase output, it must have at least 100 confirmations;

9. reject if sum of input < sum of output;

10. 0 < output < 21,000,000 and 0 < input < 21,000,000;

11. number of signature operations < signature operation limit; and

12. scriptSig can only push number on the stack while scriptPubkey must match isStandard.

These criteria change over time, to prevent certain types of attacks or to allow faster / slower block creation. It helps to prevent double spending, make sure that there are only new bitcoins created when it is a coinbase transaction and it makes sure that the combined scripts are actually valid. When a transaction is accepted, it will be added to the memory pool[102] until it can be mined with other transactions in a candidate block. Which transactions that will be included in the next block is based on the age of the unspent transaction output (UTXO)[103] that is being spent in their inputs.[104] How can one determine the priority of a transaction? That will be determined based on the sum of the value and age of the inputs divided by the total size of the transaction. There is a specific transaction space for high priority transactions of 50 kb.[105] Here the high priority transactions can be added, even if they carry no transaction fees.[106] The first transaction that will be added to the block is the generation transaction (or coinbase transaction), which is virtually a payment for the mining effort. This transaction has not UTXO as input but has the so-called "coinbase" as input which is used by the miner to collect his reward. Good to know is that the miner cannot spend this until there are 100 block confirmations in the blockchain.

If we now return to the chain itself, we can see that there are three types of chains being maintained by the network nodes. There are the main chain blocks, the side chain blocks, and the orphan blocks. The main chain is the one with the highest associated cumulative difficulty. The side

---

[102] A transaction is valid for perpetuity but a memory pool is a transient, non-persistent form of storage. So, imagine that all the nodes that received a transaction that is not yet in a block is being reset or restarted. The transaction will be wiped of the memory pool. The solution will be that the consequent wallet retransmits the transaction or reconstructs it with higher fees.

[103] The age of the UTXO is the number of blocks that have elapsed since the UTXO was recorded.

[104] It can even be the case that prioritized transactions can be sent without any fees if there is enough space in the candidate block.

[105] A high-priority transaction.

[106] Although some mining nodes choose to ignore these transactions.

chains are created through forks of the main chain, while orphan blocks are those that are kept in the orphan pool until its parent is discovered so that it can be added to the chain.

Table 2.10: Generation transaction		
Field	**Description**	**Size**
Transaction hash	All bits are 0	32 bytes
Output index	All bits are 1	4 bytes
Height	Block height (required since BIP34)	4 bytes (varies)
Coinbase data size	Length of the coinbase data	Varies
Coinbase data[107]	Arbitrary data	Varies
Sequence number	0xFFFFFFFF	4 bytes

Below you can see how the transaction actually looks like when it has been mined in a block. You see that the structure has changed and that different information is given than when the transaction has originally been transmitted throughout the network.

Table 2.11: Transaction inside a block[108]		
Field	**Description**	**Size**
Version	Version number	4 bytes
Flag	Optional (always 0001)	Optional 2 byte
In-counter (tx_int count)	Number of inputs	1–9 bytes
Inputs (tx_in)	The actual inputs	Varies
Out-counter (tx_out count)	Number of outputs	1–9 bytes
Outputs (tx_out)	The actual outputs	Varies
Witnesses	List of witnesses (only if there is a flag)	Varies
Lock_time	If ≠ 0 and sequence < 0xFFFFFFFF: block height or timestamp when final	4 bytes

Interesting here is the mentioning of inputs and outputs. The inputs of a transaction actually refer to the output of a previous transaction. Each transaction input is linked to a previous output because you can simply not spend what you have never received in the first place. So, the network wants to know two things: (1) where are these bitcoins coming from and (2) are they actually yours to spend? Similarly, the outputs provide this information for the transactions that will follow. Lock time is used to delay a transaction. Based on this value, a transaction can only be added to a block

---

[107] It was in here that Satoshi Nakamoto hid his secret message. However, the beginning of this field is no longer arbitrary with BIP00034 which states that version 2 blocks start the coinbase data with the block height index as a script push operation.

[108] (December 13, 2019). Transaction. *Github – Bitcoin*. https://en.bitcoin.it/wiki/Transaction. Accessed December 26, 2019.

unless a certain block height (or Unix time) has been reached. We have mentioned several times the pubkey script or sciptPubKey. This is a script that is included in the outputs and sets the conditions on when these satoshis can be spent. The data that is needed to fulfill these conditions can be provided by making use of the signature script (or scriptSig). Several opcodes can be used in the scriptPubKey when it comes down to our transactions. Below we give a short overview of the opcodes than can be used in the script.

- **OP_TRUE / OP_1:** pushes values 1–16 up the stack depending on OP_1 to OP_16;

- **OP_CHECKSIG:** consumes a signature and a full public key. It checks if the transaction data specified by the SIGHASH flag was converted into the signature by the same ECDSA private key. If so, TRUE is pushed on the stack, otherwise FALSE;

- **OP_DUP:** pushes a copy of the top item on the stack;

- **OP_HASH160:** computes RIPEMD160(SHA256()) of the topmost item on the stack;

- **OP_EQUAL:** checks if the top 2 items are equal and pushes TRUE or FALSE on the stack;

- **OP_VERIFY:** if the topmost item is 0 (FALSE), it terminates the script in failure;

- **OP_EQUALVERIFY:** runs OP_EQUAL and OP_VERIFY in sequence;

- **OP_CHECKMULTISIG:** consumes the value (n) at the top of the stack, consumes that many of the next stack levels (public keys), consumes the value (m) now at the top of the stack, and consumes that many of the next values (signatures) plus one extra value; and

- **OP_RETURN:** terminates the script in failure.

A final point that we should consider with Bitcoin transactions is the concept of "confirmations." When you perform a transaction, this transaction has to be confirmed over time. This confirmation can only happen when the transaction that you just performed, is propagated throughout the network and accepted in one of the blocks. In the case of the Bitcoin network, this happens, as you know, every 10 min. However, to be entirely sure, it is better to wait until a couple of blocks have been mined on top of the block containing your transaction. In actuality, if a vendor wants to be sure that a transactions is really valid, he should wait about an hour (5–6 blocks on top of the block containing your transaction) before he could really accept the transaction as confirmed and valid. For small transactions this process might be a little bit over the top but for payments that have some significance this process can prevent several types of attacks and attempts on double spending. In the world of financial services this is also known as "clearing."

2.5   BITCOIN SIGNING AND VERIFICATION

We already went through the chaining of blocks to create the blockchain and we have seen how transactions are structured and shared by the peer nodes. But how do we sign transactions and how are these being verified? Bitcoin makes use of the Elliptic Curve Digital Signature Algorithm, or ECDSA for short. With a signature we sign a scalar for G (we defined this earlier for Bitcoin in the section on elliptical curve cryptography). Signing with ECDSA comes down to the use of the discrete log problem. Signing can only work if you either know the private key or if you are able to break the discrete log problem. We make use of a signature hash to create a fingerprint of the input data (a hash algorithm will always give us a fixed-size output, no matter the size of the input). This to make sure that we sign something that is exactly 256 bits. The equation behind the signing process looks like the following:

=> the relation between the private key e and the public key P => $eG = P$;

=> a target k which has a similar relationship => $kG = R$;

=> the discrete log problem implies => $uG + vP = kG$ with k a random 256-bit integer and $u,v \neq 0$ chosen by the signer. We also know P and G;

=> $u = z/s$ and $v = r/s$;

=> $s = (z + r)/k$ where r is the x-coordinate of R and z the signature hash.

It is this final equation that gives us the final signature algorithm that is being used in ECDSA. Verification is easier (which is also the entire purpose within the blockchain network):

⇨ we know the hash z, the public key P of the signer, the x-coordinate of R, and the signature s;

⇨ $u = z/s$ and $v = r/s$;

⇨ $uG + vP = R$;

⇨ we know the signature to be valid if the x-coordinate of R and r are equal.

Distinguished Encoding Rules (DER) are used for the encoding of the r and s values that are used to perform the actual signing in a single byte stream. Verification is performed on several levels. Transactions have both inputs and outputs, where the inputs reference the previous transactions and the outputs determine the new owners of the bitcoins. Each of the input needs to have the signature that unlocks the coins from the prior transaction.[109] Most of the time this is a single transaction but there are also other possibilities. It is possible to specify a script in which two ECDSA signatures are necessary or two-out-of-three schemes are also possible. The script opcode that verifies if the signature of a transaction input is valid is called OP_CHECKSIG. De-

---

[109] (December 26, 2018). Protocol documentation. *Github – Bitcoin*. https://en.bitcoin.it/wiki/Protocol_documentation#Signatures. Accessed November 20, 2019.

pending on the script that is being used, OP_CHECKMULTISIG is also possible together with OP_CHECKSIGVERIFY and OP_CHECKMULTISIGVERIFY. Another requirement is that the sum of the inputs must be equal or greater than the output. The amount extra is considered to be transaction fee for the miner of the transaction. Some transactions also make use of locktime and transactions are marked as invalid if the specified time has not yet passed. This is the OP_CHECK-LOCKTIMEVERIFY opcode. To complete the story, digital signatures are applied to transactions. There is the SIGHASH flag that indicates which part of the transaction data is included in the hash signed by the private key. This SIGHASH flag is applied to every signature and consists out of a single byte (containing either ALL, NONE, SINGLE, or ANYONECANPAY). Depending on the type of transaction you are willing to create, a different flag is needed to indicate this to the other participants that are making use of the transaction type.

Flag	Description	Value
ALL	Signature for all inputs and outputs	0x01
NONE	Signature for all inputs	0x02
SINGLE	Signature for all inputs and one output which has the same index as the signed input	0x03
ALL|ANYONECANPAY	Signature for one input and all outputs	0x81
NONE|ANYONECANPAY	Signature for one input and no outputs	0x82
SINGLE|ANYONECANPAY	Signature for one input and all outputs with the same index	0x83

One should keep in mind that a transaction can contain several inputs that each can have a different SIGHASH flag. On top of that, the unlocking scripts can contain a signature that lead to different SIGHASH flags and lead to different parts of the transaction being committed. The SIGHASH will eventually sign the nLocktime field and itself is appended to the transaction as well. This means that it will also be hashed and therefore cannot be changed by any other participant in the network, providing the security necessary to ensure that transactions remain how they were defined when they were send out by the signing party.

2.6   UTXO

UTXO is a term we have seen earlier in this chapter. It is the basis of transactions taking place in the network as they are used to balance the ledger.[110] From the second a transaction is confirmed, the coins are removed from the UTXO database and the transaction is recorded on the ledger. To understand the way one can spend bitcoin, we should understand that multiple fractions of bitcoin

---

[110] Asolo, B. (December 20, 2018). Bitcoin's UTXO set explained. *Mycryptopedia*. https://www.mycryptopedia.com/bitcoin-utxo-unspent-transaction-output-set-explained/. Accessed November 28, 2019.

can be retrieved by an algorithm to fulfill a transaction. The changes made to each of these fractions used is sent to the UTXO database. Of course, one needs to take into account the transaction fees that one has to pay to execute a transaction. These fees are removed from the UTXO so that this amount will always be lower than the amount you originally have send out. So, while we have the idea in our heads that transactions send bitcoins from one wallet to another, they actually move from transaction to transaction. As explained before, each transaction has an input which is the output from a previous transaction. As the transaction can in case create several outputs, the outputs themselves can only be used once, otherwise you would be attempting to double spend. That is why we have unique transaction identifiers (TXIDs). In the world of Bitcoin you have two types of outputs: spent and unspent. That is why UTXOs are so important, because the money in your wallet isn't really in your wallet. It is in the UTXO database.

To explain with a practical example: imagine that there are two participants A and B and participant A wants to send 0.80 BTC to participant B. He will also have to pay a transaction fee of 0.20 BTC. To perform the transaction he will make use of a set of UTXOs as inputs. (In the image they are all the same size while in reality this will often not be the case.) These unspent UTXOs can be found in a global database that is kept and updated by all full (and up to date) nodes of the Bitcoin network. When UTXOs are used, the set is reduced, while new unspent output is created, enlarging the global UTXO set. This transaction, and all other transactions, reference both the address of the current "owner" of bitcoins (and a change address which can be the same or different from the input address) and the address of the new owner. Any "change" from the transaction can be sent to the change address. This is because UTXOs cannot be split before you use them as input.

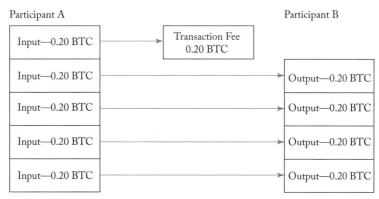

Figure 2.2: UTXOs.

There is also the possibility to create unspendable outputs from transactions, by making use of the Bitcoin scripting language. Specific applications can be developed by making use of this scripting language called "smart contracts" (explained in detail later on). Sometimes these applications can develop output that cannot be spent and when the RETURN operator is used, these are no longer stored in the UTXO database (otherwise it is stored in there, even though you can no longer spend it, making it ever more expensive to run a full node). Here it is also interesting to note that Satoshi is the smallest indivisible unit of the bitcoin currency. Clearly called after the inventor of the Bitcoin network, a lot of transactions and transaction fees happen in Satoshi instead of in bitcoins.

Satoshi	Bitcoin	Alternative name
1	0.00000001	
10	0.0000001	
100	0.000001	1 Bit/µBTC/you-bit
1,000	0.00001	
10,000	0.0001	
100,000	0.001	1 mBTC/em-bit
1,000,000	0.01	1 cBTC/bitcent
10,000,000	0.1	
100,000,000	1.0	

One final aspect is would like to note is the existence of "bitcoin dust," which is an output that requires a transaction fee that is greater than 1/3 of that outputs value. This is a situation that is very uninteresting for involved parties as the costs are extensive when we look at the actual value. Combining several of these outputs into a more valuable output can create a more interesting transaction for all participants involved.

2.6.1   TIMELOCKS

Another interesting implementation is the one called "timelocks." These allow for the restriction of transactions or the use of outputs until a certain point of time. It has been there since the very beginning with the ,nLocktime-field but new features were later introduced: CHECKLOCKTI-MEVERIFY and CHECKSEQUENCEVERIFY. The first field helps define the earliest time that a transaction is valid and can actually be processed and validated. The value given to this field should be below 500,000,000 and is interpreted as a block height. This means that the transaction can only become valid when the blockchain reaches this block height. Above 500 million it is interpreted as a Unix Epoch timestamp and the transaction is only valid when the given time is reached. Problem is that nLocktime can give rise to double spending. As you can send out a transaction with a locktime and another without one. BIP-65 came to the rescue with the CHECKLOCKTIMEVERIFY implementation or "CLTV." The difference with nLocktime is that CLTV really is an output-based timelock while the nLocktime-field is a transaction-level timelock. CLTV isn't meant to replace nLocktime, but rather prevents the spending of the transaction value UTXO until a time where nLocktime is set to a greater or equal value. The previous examples are absolute timelocks while there are also possible relative timelocks that can be used in the Bitcoin network. These timelocks depend on elapsed time based on the confirmation of the output in the blockchain. This can be interesting to keep a set of transactions off-chain and is used in lightening networks and payment channels (see later in this section). A first field that can be used for this is called "nSequence" and is the transaction-level timelock, while the script-level timelock is called "CHECKSEQUENCE-VERIFY" or "CSV." The nSequence-field is standard set to 0xFFFFFFFF when the transaction doesn't make use of timelocks. If CLTV or nLocktime is being used, it should be set to a value less than 2^{31}. BIP-68 introduced some uses for this field as a value less than 2^{31} is now being interpreted as a relative timelock transaction. Each of the inputs can have a different relative timelock so the transaction only becomes valid until the last input has become valid. The nSequence can neither be specified in the number of blocks or seconds. To differentiate, there is a type-flag in the 23rd least-significant bit. If it is set, the value is interpreted as a multiple of 512, otherwise it is interpreted as a number of blocks. The CSV is linked to nSequence, just as CLTV is linked to nLocktime. So, what is a possible problem that you can spot here? All of these implementations deal with time. But time in a decentralized P2P network means something different than in centralized systems where a server can determine the time for all clients. In the world of Bitcoin, each participant has his own interpretation of time. On top of that, there is network latency that puts even more strain on the interpretation of time throughout the network. This can create difficulty for both transaction being interpreted, timestamps and of course timelocks. Miners are incentivized to lie about the timestamp they set on the blocks they mine if they contain timelocked transactions, to earn even more fees (this is also known as "fee sniping").[111] This incentive was removed with BIP-113 with the introduction of "Median-Time-Past." This timestamp is calculated by taking the last 11 blocks and calculates the median time. This becomes the consensus time in the network and is eventually used for the timelocked transactions. The median-time-past will approximately be 1 hr behind the "wall-clock time."

2.7   BITCOIN SERIALIZATION

Serialization is used to communicate all the information over the Bitcoin network. Several formats have been used in the past to make use of serialization. Today there is a specific Bitcoin standard (based on SegWit or non-SegWit). In the past it made use of the uncompressed SEC format, also known as "Standards for Efficient cryptography," to do so. The uncompressed SEC format for a point $P = (x,y)$ is generated by following these steps:

⇨ prefix: 0x04;

⇨ append x-coordinate 32 bytes as a big-endian integer; and

---

[111] Fee sniping is not lucrative now as the block reward is high enough but in some point in the future the transaction fees will be high enough to attempt and steal them. This is why nLoctime is set to "current block + 1" and nSequence to 0xFFFFFFFE so that locktime is set to the next block and fee sniping becomes impossible.

⇨ append y-coordinate 32-bytes as a big-endian integer.

For the compressed SEC format, we use the following procedure for a point $P = (x,y)$:

⇨ prefix: 0×02 or 0×03 (if y is even the former, otherwise the latter) and

⇨ append x-coordinate 32-bytes as a big-endian[112] integer.

From the procedure we can clearly see that the compressed format is 33 bytes opposed to the 65 bytes of the uncompressed format. For the serialization of signatures we needed to use a different format. We cannot simply compress the signature because of what we have seen before (the discrete log problem) we cannot derive s from r. Therefore, we make use of DER or "Distinguished Encoding Rules." We can find this standard in the OpenSSL library. It has the following format:

⇨ 0×30 (start);

⇨ encoding of the length of the signature;

⇨ 0×02 (marker byte);

⇨ encode r (big-endian) and prepend the resulting length of r;

⇨ 0×02 (marker byte); and

⇨ encode s (big-endian) and prepend the resulting length of s.

The current format that is used since SegWit and BIP 141, while before the segregated witness update, Bitcoin was making use of "raw format." The raw transaction format, which was in use since October 2014, had the following format:

Table 2.12: Raw format			
Name	**Data type**	**Description**	**Bytes**
Version	Uint32_t	Transaction version nr	4
Tx_in count	compactSize uint	Nr of inputs in transaction	Varies
Tx_in	TxIn	Transaction inputs	Varies
Tx_out count	compactSize uint	Nr of outputs in transaction	Varies
Tx_out	txOut	Transaction outputs	Varies
Lock_time	Uint32_t	A time or block number	4

We can see in Table 2.12 a data type called "compactSize Unsigned Integers." This data type is used in the raw transaction format to indicate the number of bytes that can be expected in the following data set that is going to pass. When you read up in documentation on the Bitcoin network, you will also find this as "var_int" or "varInt" as it is a variable length integer. Table 2.13 below indicates how the encoding scheme works, with numbers 0–252 looking like regular unsigned integers.

---

[112] Big-endian means that the most important byte is written first; little-endian means the least important byte is written first.

Table 2.13: The encoding scheme		
Value	**Bytes Used**	**Format**
>= 0 && <= 252	1	Uint8_t
>= 253 && <= 0xffff	3	0xfd + number as uint16_t
>= 0x10000 && <= 0xffffffff	5	0xfe + number as uint32_t
>= 0x100000000 && 0xffffffffffffffff	9	0xff + number as uint64_t

Finally there is the "wallet import format" or WIF which is used for our private keys. It has the following procedure:

⇨ prefix 0×80 (mainnet);

⇨ secret encoded (32-byte big-endian);

⇨ if public key is compressed SEC, add 0×01 as a suffix;

⇨ take a copy of the previous three combined, SHA256, and take the first 4 bytes; and

⇨ combine all and translate to BASE58.

2.8   BITCOIN SCRIPT

The Bitcoin network makes use of script to lock and unlock coins.[113] This scripting language is a programming language that processes one command, containing either elements or operations, at a time. As you might have guessed, the elements are data while the operations are functions that are performed on the stack, i.e., OP_HASH160 will perform a SHA256 followed by a ripemd160. These commands are known as "opcodes" and make up the programming language of the Bitcoin network. It is a living language as opcodes can be added or removed over time. Most of the time opcodes are removed or limited to make sure that the danger to the network is reduced as much as possible. It is a transaction network with a native coin and it was never the intention of the network to be anything else. So with Bitcoin script we have a push and pop style programming language that makes use of a stack and makes use of a last-in first-out data structure. Typically, "push" can be used to add an element to the top of the stack, while "pop" removes the top element. During the parsing of the script it is determined whether it is an operation or an element based on the byte (when the value lies between 0×01 and 0×4b, we know the next n bytes are an element). Whenever we would like to perform an evaluation, we combine the ScriptPubKey and ScriptSig fields, which represent the locking and unlocking mechanisms. When we were talking about blockchain addresses in the first section, we already talked about a couple of standard scripts. A couple of these are p2pk (pay to public key), p2pkh (pay to public key hash), p2sh (pay to script hash), p2wpkh (pay to witness pubkey hash), and p2wsh (pay to witness script hash). You are of course not limited to these specific standard scripts and can build your own locking and unlocking scripts. For those of you that aren't

[113] (June 27, 2020) Script. Bitcoin Wiki. https://en.bitcoin.it/wiki/Script. Accessed July 7, 2020.

following the story so far, you should imagine the world of scripting as a stack of information on top of each other. It is up to the scripts you use to make changes to this stack. An example is the OP_CHECKSIG instruction. This instruction will combine the signature with a full public key and will push TRUE on top of the stack in case the signature and the public key are generated from the same private key, else there will be a FALSE. OP_CHECKMULTISIG does something similar but with multiple signatures and public keys. Several sets of opcodes can be defined: the constants, opcodes used for flow control, opcodes to perform specific stack operations, there is still 1 splice opcode called OP_SIZE, 2 bitwise logic operators (OP_EQUAL and OP_EQUALVER-IFY), some arithmetic commands, hashing and signing opcodes, locktime opcodes, reserved and pseudo words. For a more conclusive overview of these instructions and Bitcoin script, I would advise you to check out the Bitcoin developer reference and the Bitcoin Wiki pages. As mentioned before, you should be aware that changes happen to these OP_CODES and that those that were once in use can perfectly be disabled (in an attempt to increase the security of the Bitcoin network). When you see these instructions, and the way of working (stack), you can understand why this is not immediately that developer friendly. It tells you clearly that the original intent of Bitcoin wasn't the creation of smart contracts (which doesn't mean that this isn't possible).

2.8.1   IVY FOR BITCOIN

Ivy[114] is a high-level programming language that was developed by Chain and allows to write smart contracts for the Bitcoin protocol.[115] At the moment it is still a prototype software but it can already be used to test software and look at how you could possibly develop new applications in the future with this (new) programming language. There is a playground available where you can test out contracts you have written. It offers you all the flexibility of the Bitcoin script as you would expect but you have some extras which make it interesting to look into this language: name variables, name clauses, static types, and familiar syntax for both functions and operators. It has some similarities to the Solidity or Vyper languages used for Ethereum smart contract development and other platforms (later discussed in more detail). You always have to specify a contract template and pass some arguments to be able to actually make use of the contract. These arguments have several types such as PublicKey, Value, and Signature. The contract also needs some contract clauses, the arguments needed to unlock a contract.

The types that are defined in Ivy for Bitcoin script are:

- **Bytes:** a string of bytes;

- **PublicKey:** an ECDSA public key;

---

[114]   Robinson, D. (2018). ivy-Bitcoin. https://docs.ivy-lang.org/bitcoin/language/IvySyntax.html. Accessed December 6, 2019.

[115]   Chain has been acquired by Lightyear, which is a Stellar-focused company, and together they form "Interstellar."

- **Signature:** an ECDSA signature;

- **Time:** block height or timestamp;

- **Duration:** number of blocks or x times 512 sec;

- **Boolean:** True/False;

- **Number:** an integer;

- **Value:** amount of bitcoins;

- **HashableType:** any type you can pass to hash functions (Bytes, PublicKey, or result hash functions);

- **Sha256(T: HashableType):** SHA256 of HashableType T;

- **Sha1(T: HashableType):** SHA1 of HashableType T; and

- **Ripemd160(T: HashableType):** RIPEMD160 of HashableType T.

The following functions have been defined in Ivy for Bitcoin script:

- **checkSig(publicKey: PublicKey, sig: Signature):** Boolean result from the check if the public key is the one that corresponds to the private key used to make the signature;

- **checkMultiSig(publicKeys: [], Sigs: []):** Boolean result from the check that each of the public keys correspond to the private keys used to create the signatures;

- **After(time: Time):** Boolean that checks if the current block height / time is after time. Uses nLockTime and CHECKLOCKTIMEVERIFY;

- **Older(duration: Duration):** Boolean that checks if the contract being spent has been on the blockchain for at least duration. This uses CHECKSEQUENCEVERIFY;

- **Sha256(preimage: (T: HashableType)):** SHA256 of preimage;

- **Sha1(preimage: (T: HashableType)):** SHA1 of preimage;

- **Ripemd160(preimage: (T: HashableType)):** RIPEMD160 of preimage;

- **Bytes(item: T):** turns item into bytestring, cannot be performed on Value or Boolean (only affects type checking);

- **Size(bytestring: bytes):** returns "number" type, the length of the bytestring; and

- **== or !=:** equality or inequality check.

As you can see, this feels a lot more familiar for most of you than the actual Bitcoin scripting language. For those of you that are interested, I would certainly advise to visit the website and use the documentation to gain deeper insight. It will help you to get a better understanding of the Bitcoin scripting language, and for those of you that are willing, it might be the start point to actually programming in Bitcoin script (mastering Bitcoin by Andreas Antonopoulos might give deeper insight in the scripting language itself). Below you can find an example in both Bitcoin script and in the Ivy programming language to give you an idea on how the higher-level Ivy programming language actually compiles to opcodes. We are going to take the LockWithPublicKeyHash example as it can tell us how the contract language translates to the Bitcoin scripting language.

```
contract LockWithPublicKeyHash(pubKeyHash: Sha256(PublicKey),
  clause spend(pubKey: PublicKey, sig: Signature) {
    verify sha256(pubKey) == pubKeyHash
    verify checkSig(pubKey, sig)
    unlock val
  }
}
```

Figure 2.3: LockWithPublicKeyHash.

You can see here clearly that you use the hash of the public key to lock the contract, while you need to provide the public key to unlock it. This contract would eventually compile to the following output:

OP_DUP OP_HASH256 <pubKeyHash> OP_EQUALVERIFY OP_CHECKSIG

The actual script would look like this:

ScriptSig: <sig> <pubKey>

ScriptPubKey: OP_DUP OP_HASH256 <pubkeyHash> OP_EQUALVERIFY OP_ CHECKSIG

Finally, there is also a JavaScript library that is in its early stages but can be used for testing.

2.9   BITCOIN MINISCRIPT

The latest innovation in Bitcoin scripting is called "Miniscript" which is a language that should allow developers to write scripts in a more structured way. It also allows statistical analysis for various actions such as spending conditions, correctness, security properties, and malleability.[116] It has currently been developed for P2WSH and PS2H-P2WQH scripts (and also adheres to its re-

---

[116]   Wuille, P., Poelstra, A., and Kanjalkar.S. (2019). Analyze a miniscript. *Blockstream*. http://bitcoin.sipa.be/Miniscript/. Accessed December 7, 2019.

source limitations as defined in the Bitcoin scripting language by either standard or consensus[117]).[118] Currently, there are Bitcoin core compatible C++ and a Rust implementation. On the dedicated website you can test and analyze Miniscripts and you can see if you have actually achieved what you wanted to. You can also find a thorough reference that you can use in the further development of your scripts. If you look through the list you can clearly see how Miniscript could make the life of many a developer easier. An example is the following Bitcoin script: "SIZE <32> EQUALVERIFY RIPEMD160 <h> EQUAL" gets reduced to "ripemd160(h)." However, not every Miniscript expression can just be compiled with another. Just as any other Bitcoin script (or even any other programming language) the expressions linked together need to make logical sense. In Miniscript there has been a "correctness type system" introduced to help developers from making mistakes. There are four types.

- **Base expressions (B):** these expressions take their input from the top of the stack. This type is used for most of the expressions and is required for the top-level expression. If the expression evaluates to true, it pushes a non-zero value to the stop of the stack, otherwise it will be zero.

- **Verify expressions (V):** these expressions take their input from the top of the stack. If the expression evaluates to true, nothing happens and the script continues. Otherwise, there will be an abort of the script.

- **Key expressions (K):** taking their input from the top of the stack, they always push a public key onto the stack which requires a signature to satisfy the expression.

- **Wrapped expressions (W):** these expressions take their input from one below the top of the stack and when it evaluates to true, push a non-zero value to the top of the stack. Otherwise, a zero value is pushed to the top of the stack.

Conversions between these expression types are also possible. You can convert a B expression in a V expression, a K into a B, and so on. The Miniscript language also comes with five types of modifiers:

- the z or "zero-arg" that consumes exactly 0 stack elements;

- the o or "one-arg" that consumes exactly 1 stack element;

- the n or "nonzero" that consumes at least 1 stack element;

---

[117] The standardness and consensus rules are built-in the Miniscript language so that you can make sure that your scripts adhere to the Bitcoin scripting rules.

[118] The design and implementation has been done by Pieter Wuille, Andrew Poelstra, and Sanket Kanjalkar at Blockstream Research.

- the *d* or "dissatisfiable" which allows an unconditional dissatisfaction to be constructed; and

- the *u* or "unit" that, when satisfied, will add a "1" on the stack.

If we keep the example we looked at previously (ripemd160(h)), we know that this is a base expression (B) and that the properties are *o*, *n*, *d*, and *u*. The developers of the Miniscript language have also built-in satisfactions and dissatisfactions for each of their scripts. For ripemd160(h), the satisfaction is a preimage while the dissatisfaction is any 32-byte vector except for the preimage. So, the correctness of an expression is based on these predefined rules within Miniscript. However, there is one important point that we need to take into consideration: malleability. The Miniscript language has been created with the Segwit update in mind. This means (as you hopefully remember) that changing certain parts of the transaction no longer breaks the validity of unconfirmed descendant transactions. Still, there might be certain undesirable effects.[119] Miniscript was designed as such that non-malleable signing is permitted, increasing the security of the transactions that can be created. The non-malleable satisfactions within the language are created by making use of a function that returns the optimal satisfaction/dissatisfaction for a certain expression or a special DONTUSE value and an optional HASSIG marker that shows if the solution contains at least one signature. The function should recursively be used over all the subexpressions in the function you are creating. You should know that certain expressions (such as the ripemd160(h)) are always malleable so that you have to use the DONTUSE value. With each check one should make sure that every possible outcome has the HASSIG marker, otherwise the (sub)-expression is malleable in one way or another. There are of course the *d*-expressions that are unconditionally dissatisfiable, meaning that there must be a non-HASSIG dissatisfaction. Here the rule is that non-HASSIG solutions must be preferred over HASSIG solutions and when there are multiple non-HASSIG solutions, none can be used.

Depending on the expressions you are using, there will be different requirements for one to make sure that their Miniscript is nonmalleable. You should always strife to make a script nonmalleable as this greatly increases security and trust for all the involved participants.

2.10   BITCOIN ADDRESSES

We already offered a short discussion on blockchain addresses. Still, it is of interest to make note of them here and the formats on how we can find them and how they are derived. As we know, we have to make use of a public key and a private key. There are several private key formats that we should take into account.

---

[119] An example given in the Miniscript documentation is that the witness can be stuffed with additional data, forcing the feerate down and eventually impacting the transactions possibility of being processed and confirmed.

Table 2.14: Private key formats

Type	Description / size	Prefix
Raw	32 bytes	None
Hex	64 hexadecimal digits	None
WIF	Base58check encoding	5
WIF-compressed	Base58check encoding with 0×01 suffix before encoding	K or L

Public keys on the other hand either exist in compressed or uncompressed formats. Depending on the format, the prefix is either 04 (uncompressed) or 02/03 (compressed). As explained before, the compressed version is the standard but some of the older clients that are still part of the network do not yet support these. To resolve the possible problems leading from this difference, when private keys are exported from a Bitcoin wallet to another wallet, the WIF is implemented differently to indicate that these keys have been used to produce compressed public keys, leading to compressed Bitcoin addresses.[120] Commonly, the P2PKH (pay to public key hash) and P2SH (pay to script hash) are used. As you can clearly see in the name, these two procedures make use of hashing. This increases the inherent security issues that you might find with classic approaches, such as the very first Bitcoin network where you had to pay to the IP-address of a peer. For the P2PKH, you might have guessed, that you have to hash your public key. This hash consists out of SHA256, followed by RIPEMD160. For the P2SH you have to use a similar hashing format but instead of your public key, you hash a so-called "redeem script." The next step in the creation of your Bitcoin address, consists out of adding a version byte. For the Bitcoin mainnet this is 0x00 for the P2PKH and 0x06 for P2SH.[121] The third step in the process is creating a copy of this hash combined with the version byte and hash this again twice with SHA256. From this result you take the first 4 bytes as a checksum to make sure that your original hash + checksum is transmitted correctly. You append this checksum to the version + hash combination. The final step is encoding the result in a BASE58check string. This is the BASE58 encoding format but with a built-in code that checks for possible errors. In practice, this leads to a checksum with four additional bytes being encoded. This checksum is later used by the software to determine the validity of the encoded data. In SegWit the BASE58 is being replaced with Bech32 because it is more user friendly (only lower-case letters and numbers). So what are P2SH functions used for currently? Most commonly, it can be used for something such as multi-signature address scripts. These scripts allow the creation of addresses that, i.e., have a maximum of three participants that can sign, of which two need to sign to approve a transaction. We can also introduce the concept of "vanity" addresses which contain a specific human-readable message. This means that the address contains certain words or numbers that you

---

[120]   Antonopoulos, A. (2017). *Mastering Bitcoin: Programming the Open Blockchain*. 2nd ed. California: O'Reilly Media.

[121]   In the Bitcoin testnet this will be 0×6F and 0×c4, respectively.

can choose. This can refer to your company or you as a person. These addresses are just as secure as any other address but the search time to find an address that contains all the letters and numbers you wish for, can take up quite an amount of time. A pattern of up to 6 characters will take about an hour or less, while 8 characters will already take up to four months and nine characters 800 years (considering you use your personal computer at home and not some advanced supercomputer).

2.10.1 ENCRYPTED PRIVATE KEYS (BIP-38)

There have been several Bitcoin Improvement Proposals to increase the security of the wallets in use. This to provide protection to the users but as it often is, there is also a dark side to some of these implementations. BIP-38 introduces the encryption for private keys. AES is used in this standard to encrypt them so that the information itself can be kept safe. These encrypted keys will always start with "6p." Current wallets often are able to recognize these encrypted keys and will ask you for your passphrase. This increases the security of the private keys themselves but can lead to other issues. If you lose your passphrase, you no longer have access to your private keys and these cannot be recovered.

2.11 BITCOIN WALLET

We can complete the explanation of the Bitcoin network itself with the wallets that are being used within the network. The usefulness of mining, signing, and verifying transactions would be rendered completely useless if there would not be something as a wallet. It is in this wallet that you store your bitcoins (currency is written "bitcoin," the network as "Bitcoin"). It is basically a combination of your Bitcoin address and your private key that makes up your wallet. There are nondeterministic wallets which indicates that each key is independently generated by making use of a random generator. We also know these wallets as JBOK wallets or "just a bunch of keys" wallets. Problem is that each address has to be backed up and used for multiple transactions, reducing privacy and security of the user. The type-0 nondeterministic wallet was introduced by the Bitcoin core implementation but shouldn't be used anymore. Compared to other implementations they take too much work backing up and use, certainly if you take privacy into account. In short, we can identify two types of wallets that we actually can use, where the first one is the type one deterministic wallet, called the single chain wallet. This wallet can only send and receive a specific cryptocurrency, in this case bitcoins. It is the simpler version as it helps to create a single series of keys from a single seed. This immediately implies that if the seed is leaked or stolen, all funds are in jeopardy. The second type of wallet is the Hierarchical Deterministic wallet or HD wallet (based on BIP-32/BIP-44). This is a solution where the wallet software can generate a pattern of public and private keys which do not require a backup and cannot easily be guessed. It is generated from a root seed that consists out of a random number that is either 128, 256, or 512 bits. This seed is inserted in a HMAC-SHA512 algorithm to generate a master private key and a master chain code. From the master private key, the master public key is generated and the chain code is used as entropy for all the child keys that are generated afterwards with the "child key derivation function" (CKD). The master key and the chain code are concatenated and form the 512-bit "(private or public) extended key."[122] The CKD makes use of the HMAC-SHA512 hashing algorithm, an index number, the chain code and the parent key are combined to generate a hash. This hash is split in two where the right half is used as the chain code for the child while the left half is added to the parent key to generate the child key. The index allows us to create up to 2^{31} children, where the children can become parents again and perform the same process. In practice, the CKD is a bit changed to increase security (if the chain code and a child private key are leaked, one could guess the master private key and breach all keys in use). There is a hardened key derivation process which makes use of the parent private key to derive the child chain code instead of the parent public key, breaking the relationship and making it impossible for attackers to derive the master/sibling keys.

It is clear that within HD wallets an infinite number of keys can be generated in a tree-like structure which can be restored by making use of a seed word. This is the set of 12 words you get when you generate such a wallet. When you log in again, the wallet asks typically for a couple of the seed words and this way is capable of regenerating the keys and with it your funds. So, it remains crucial that you store the seed key. BIP-44 allows for wallets that introduce both multiple accounts but also multiple currencies! An example of the HD paths and coins you can find in Table 2.15.

Table 2.15: HD paths and coins

HD path	Decription
m/44'/0'	Bitcoin
m/44'/1'	Bitcoin testnet
m/44/2'	Litecoin

These HD wallets are improved by making use of BIP-39 that allows for a standardized way of creating seeds from a specific sequence of English words. This not only enhances the inner working of the wallet but also has a positive impact on the user experience. The number of words that are used, usually are a sequence of 12–14 words. How does the wallet actually create this link? Well, it comes down to this: a random sequence of 128–256 bits are generated, after which a checksum is generated by taking the first (entropy-length/32) bits of its SHA256 hash. The checksum is added to the end of the sequence, and next the entire result is split in 11-bit segments. These segments are mapped to a dictionary of 2,048 words. These words from the mnemonic words that you will have to store. These words are hence used to generate a seed by making use of both the words, a salt and the PBKDF2 function. This seed can either be a constant string within the software or

---

[122] They are encoded in BASE58Check format and use the prefix "xprv" or "xpub."

a passphrase. This passphrase can provide extra security by making the original mnemonic words useless without the passphrase, and the creation of a duress wallet which can distract attackers from the actual wallet. Another BIP that helps improve the user experience and usability of wallets are the BIP-43 that allows for multipurpose wallets. If you are looking for a specific wallet client, please check out https://bitcoin.org/en/choose-your-wallet where you can find an overview with some amazing wallets to use and create. For each operating system (mobile or desktop), you will find sufficient possibilities.

2.12   SIMPLIFIED PAYMENT VERIFICATION

SPV is a technology which allows you to validate transactions without taking into account the transactions of other participants. It makes sure that your transactions are included within a block that can be added to the chain and it provides confirmations of blocks being added to the chain. Why can this be interesting? Do you have an idea what the current size is of the Bitcoin blockchain? As of April, 2019, it is a whopping 210 GB! So, if you wish to store the entire blockchain database, you need quite some storage space! If you are not interested in supporting the network but only in its feature as a cryptocurrency, this is quite a dilemma. It is not as if you are going to carry around an external hard drive just so that you can keep on making payments. It is here that the Merkle root, one of those pesky elements we explained in the very beginning, comes in handy. It can be used as a proof of inclusion so that the client can verify that a transaction was included in a block without having to know every transaction that at one point has taken place. A clever reader should be able to identify an issue. If you work with a client that has SPV as a way of working, it means it is dependent on a full node to receive information concerning the block so that the light client can verify the transaction with the Merkle root. If I have a full node and you have the light client, I can try to deceive you. How? You have no assurance that the block I am presenting is actually in the blockchain itself. This way you have no way of knowing for sure that I am speaking the truth. The only way around this is by connecting with many other full nodes, as they can display the truth. Still, it is advisable to work with full nodes if you are making large payments. If the payment is high enough, nodes might always have the incentive to try and deceive you. A final aspect that I would like to address here before we go on is a change brought by BIP37: the use of bloom filters. I briefly explained what a bloom filter is at the beginning of the section but here we can see its use. Bloom filters are used for network communication within the Bitcoin network. The network makes use of the murmur3 hash function, which is very fast but not cryptographically secure.[123] It is used in light clients so that these can solely focus on transactions of interest for the wallet. It makes use

---

[123] For those among you that have more in-depth knowledge of Bloom filters: it is indeed more efficient to use several hashing algorithms to reduce the space needed by the bit field. Bitcoin just uses the same hashing algorithm but changes the seed, leading to the same efficiency.

of filterload (set the Bloom filter), filteradd (add data element to the filter), and filterclear (remove the filter) messages.

2.12.1 SPV WALLET CLIENT

There are several SPV wallet client implementations but wallets making use of centralized API servers are far more popular. Currently you can easily make use of BRD wallet, Electrum, Bitpay, and several other implementations. There is the native Bitcoin SPV client and the modified SPV wallet client for Bitcoin. There are several Bitcoin native SPV wallets available and in development on Github.[124] To this day it is not advisable to use these in a production environment as they are still under development. On the other hand, there are the modified SPV wallet clients which make use of other developments, such as OpenBazaar, the lightening network, and BTCD.

2.13   SEGREGATED WITNESS

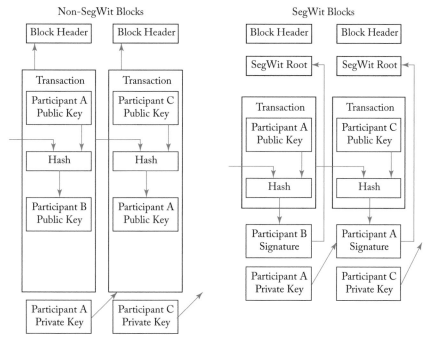

Figure 2.4: Non-SegWit vs. SegWit.

We have already mentioned it several times in this section so far—witnesses, SegWit, and segregated witness—but what is it? We mentioned it already a couple of times before that this was a soft fork in the Bitcoin network. An interesting option that was added by the SegWit soft fork is

---

124   For example, on https://github.com/keeshux/bitcoinspv.

the block size increase. As we have seen before, the maximum size of a block is 1 MB and currently this size, combined with the block time is simply not enough to handle the enormous amount of transactions that is being processed every day.[125] The solution that is being put forward is that the signature data is moved to an extended block so that space can be freed up in the original block. This is certainly interesting if you know that signature data comprises about 65% of a regular, standard block. By making use of this technique, the block size increases to 4 MB. Next, uncompressed SEC pubkeys are no longer allowed, which will also help to save space.

Another one of the things it introduced was the pay to witness public key hash or p2wpkh.[126] The only difference with p2pkh is that the location of the data for the ScriptSig is now in the witness field to prevent transaction malleability—here the unique id of a transaction is changed and this is possible because the digital signature in the ScriptSig can be modified and when one modifies this information, the unique ID changes as well.[127] With SegWit and the new location of the data, this is omitted when creating the unique ID and therefore the ID no longer changes when an attacker starts modifying the data. The witness can consist out of the digital signature but it can basically contain any condition that is necessary to unlock the UTXOs and make them available for spending. The introduction of the segregated witness scripts brings also the "script version" number which tells what type of script we are dealing with (similar to transactions and blocks). The problem with p2wpkh is that it is not backward compatible with older wallet technologies as they cannot send payments to p2wpkh ScriptPubKeys because of the change from BASE58 to Bech32. To address this problem, p2sh-p2wpkh saw the light of day where the new p2wpkh is wrapped inside the p2sh. So, it looks like a normal p2sh address but inside you can find the ScriptPubKey of p2wpkh. Even though p2wpkh was able to get rid of the issue of transaction malleability, we also needed a different way of working if multisignature technology was still to be included. For this pay to witness script hash or p2wsh was created, which is the same as p2sh, only with the ScriptSig data in the witness field. Similarly to the example of p2wpkh, there were problems when it came to older wallets, so p2sh-p2wsh was invented to circumvent this problem. A final problem addressed by SegWit is the quadratic hashing problem. For the verification of each signature, the amount of data hashing is proportional to the size of the transaction.[128] You can easily see that growing data will result in longer time necessary to verify the transaction. A second problem that comes out of the "current" signature process is that the algorithm makes no use of the bitcoin spent by the input. If you have a "cold" wallet, this becomes a problem because it is impossible to calculate the exact amount spent and the transaction fee applicable. This can only be solved by acquiring the entire

[125] Asolo, B. (November 1, 2018). What is segregated witness? *Myencryptopedia*. https://www.mycryptopedia.com/what-is-segregated-witness/. Accessed December 24, 2019.

[126] BIP0141 and BIP0143

[127] As long as the transaction isn't mined in a block yet—after this the ID is immutable.

[128] Stepanov, H. (July 1, 2019). bip-0143. *Github – Bitcoin*. https://github.com/bitcoin/bips/blob/master/bip-0143.mediawiki. Accessed December 28, 2019.

transaction, which again might form a problem in itself. If the input value of the transaction would be part of the signature, the cold wallet would no longer have a problem. Why? If the value that is provided is wrong and the transaction is signed, the signature becomes invalid and the transaction will simply not happen. Therefore, a new algorithm is used by SegWit to stop both of these problems. A new way of serialization is used with a new transaction digest algorithm (please check out BIP-143 for all the details).

2.14   BITCOIN IMPROVEMENT PROPOSALS[129]

Bitcoin Improvement Proposals or BIPs are design documents which provide information to the supporting community. When refer to the BIP purpose and guidelines (or BIP0001) we can identify three types of BIP.

1. A Standards Track BIP describes a change that will affect most or all Bitcoin implementations. This can be a change in protocol, block criteria, transaction validity rules, or any other change that might affect applications using Bitcoin.

2. The Information BIP provides guidance, and describes a design issue or offers information to the community. These BIPs are completely voluntarily and members of the community are free to follow the advice of a BIP or completely ignore it.

3. Finally, there is also the process BIP which describes a process surrounding Bitcoin. They are very similar to standards track BIPs but differ in that sense that where the standards track focuses on the Bitcoin protocol itself, the process BIP focuses on the processes surrounding Bitcoin.

If you are prompted to learn more about the BIPs already in place or how you can make your own suggestions, do not hesitate to visit the Github page!

2.15   SCHNORR SIGNATURES

A draft BIP which might improve the Bitcoin network even more in the future is one that focuses on the use of Schnorr signatures.[130] It was invented by Claus-Peter Schnorr while he was working at the University of Frankfurt in the 1980s. We have learned that signatures make up a significant part of the data that is being transmitted over the network. SegWit already gave a first push to a more scalable way of working. Schnorr signatures could be the second step in this process as it is security proof and non-malleable. Furthermore, one could move away from DER encoding for

---

129   BIPS can be found on https://github.com/bitcoin/bips.
130   Asolo, B. (February 16, 2019). Bitcoin Schnorr signatures explained. *Mycryptopedia*. https://www.mycryptopedia.com/bitcoin-schnorr-signatures-explained/. Accessed November 17, 2019.

signatures and even start with batch verification. Currently, as we have seen before, a transaction is constructed out of a batch of inputs coming from earlier transaction outputs. With Schnorr this would no longer be necessary and one signature could be used for all inputs. This could lead to an increase of the capacity of the entire Bitcoin network of almost 25%! Another advantage that is given by making use of this technique is the use of an interactive scheme (i.e., MuSig) where participants can produce one signature where they jointly signed for. These are all clear benefits over the current ECDSA signature. Currently, P2SH is being used which is a script smart contract and is therefore not really efficient. Key aggregation with Schnorr signatures would lead to less footprint, lower transaction costs, improved bandwidth, and more privacy for the participants. Key aggregation let multisigs become indistinguishable from other regular transactions. If and when the implementation would actually take place, the OP_CHECKSIG and OP_CHECKMULTISIG opcodes would be retired in favor of a new class of opcodes called OP_CHECKDLS. This will lead to the disappearance of multisigs in favor of musigs, at least for those participants that follow the soft fork with the Schnorr signature implementation. The current work on the implementation of the musig scheme by Blockstream can be followed on their Github page.[131] A first semi-formal proposal for the activation of Schnorr signatures has been made by the end of 2018 on the Bitcoin mailing list.[132] However, it will probably take some more months and perhaps even years of testing before the next soft fork also actually goes live.

2.16 TAPROOT, G'ROOT, AND GRAFTROOT

Taproot follows in the idea of Schnorr signatures to increase the privacy of Bitcoin transactions but on top of that it would allow for smart contract flexibility. Scripts would no longer be distinguishable from other transactions on the blockchain. As we mentioned already several times, the P2SH is currently used, which allows to lock bitcoins and only unlock them based on certain conditions. These conditions can be tailored to the needs of the participants and can create complex schemes. The P2SH allows these conditions to be hidden from the public, or at least at first (the script is included as a hash). When the owner spends the coins, the script and solution are revealed. The initial hash can be used to check if this is the script and can check what the requirements for unlocking were.[133] This is both heavy on data and not really private. Several solutions have been presented for this in the past. MAST, or "Merkelized Abstract Syntax Tree," was one of these proposals. All conditions are hashed in a Merkle tree and the root is used to lock up the coins. If any of the data

[131] Davies, J. (January, 2019). secp256k1. *Github – ElementsProject*. https://github.com/ElementsProject/secp256k1-zkp/tree/secp256k1-zkp/src/modules/musig?source=post_page. Accessed January 4, 2020.

[132] Towns, A. (December 14, 2018). Schnorr and taproot (etc) upgrade. *Linux Foundation*. https://lists.linuxfoundation.org/pipermail/bitcoin-dev/2018-December/016556.html?source=post_page. Accessed January 8, 2020.

[133] Van Wirdum, A. (January 24, 2019). Taproot is coming: What it is, and how it will benefit bitcoin. *Bitcoin Magazine*. https://bitcoinmagazine.com/articles/taproot-coming-what-it-and-how-it-will-benefit-bitcoin. Accessed October 2, 2019.

is revealed, it can be verified by making use of the Merkle root and path, without revealing the other data. Specifics of the MAST proposal made by Gregory Maxwell can be found online (BIP-114 and BIP-117).[134] This still reveals data and the combination of Schnorr signatures and Taproot will provide even better security. Taproot looks like MAST but always includes a condition where all participants can cooperate to simply send the funds, called the "cooperative close." It actually makes use of the special case of a top-level threshold signature or arbitrary conditions, which becomes indistinguishable from a one-party signature, by making use of a special delegating CHECKSIG. If you combine this with Schnorr signatures, you could make the transaction look like any other one. All public keys involved in the transaction could be aggregated together in a "threshold public key" and all participants their signatures in a "threshold signature." Another addition would be that all other possible ways of spending the bitcoins (so all other non-cooperative outcomes except for the cooperative close) would be combined in a different script. This script is hashed and used to adapt the threshold public key which would also have an influence on the signature. So only if the cooperative close wouldn't be possible, the threshold public key will be revealed for what it really is. There is also an implementation which is called generalized Taproot or G'root.[135] This is recursive taproot by making use of Pedersen commitments. This is a useful implementation when one makes use of additional conditions. If you do this, you could either start spending after the transaction is signed directly, if some extra conditions are satisfied or when the two points are revealed (in a Pedersen commitment, explained earlier) which actually satisfy that condition. This improves the privacy, as you can keep initially the lower layers of scripts hidden if you don't need them and you don't reveal the conditions corresponding with other keys, only the ones corresponding to the key you are actually spending with.[136] There is also the idea of Graftroot. This implementation wants to focus on a limitation within the idea of taproot. Namely, that it only provides one real alternative. Even if you could create a tree of taproots, they would offer less privacy than a single level.[137] Graftroot tries to solve this by, once again, letting participants establish a threshold key, with optionally a taproot alternative. They can then delegate their ability to sign to a script by signing that script, and only that script, with their taproot key and sharing delegation with whomever they want and choose. When the time comes to spend the coin and all signers aren't available, they can use the script and the redeeming party can satisfy the requirements of the script, combining this with the signer's signature of the script. Using this scheme, an unlimited number of alternatives can be pro-

---

[134]   Maxwell, G. (January 23, 2018). Taproot: Privacy preserving switchable scripting. *Linux Foundation*. https://lists.linuxfoundation.org/pipermail/bitcoin-dev/2018-January/015614.html. Accessed October 4, 2019.

[135]   Original proposal was the name "MAST-ended sc'roots" but as a joke toward the Mimblewimble folks with their Harry Potter references, Anthony Towns went for G'roas as a stab at Marvel.

[136]   Towns, A. (July 13, 2018). Generalised taproot. *Linux Foundation*. https://lists.linuxfoundation.org/pipermail/bitcoin-dev/2018-July/016249.html. Accessed October 10, 2019.

[137]   Maxwell, G. (February 5, 2018). Graftroo: Private and efficient surrogate scripts under the taproot assumption. *Linux Foundation*. https://lists.linuxfoundation.org/pipermail/bitcoin-dev/2018-February/015700.html. Accessed October 24, 2019.

vided, which are all executed with equal efficiency to a single alternative and their number is hidden without overhead. The current idea is that Schnorr signatures, taproot, and mast should be implemented first, after which graftroot, cross-input aggregation, and G'root could be implemented. This would increase both privacy and efficiency while still keeping the transactions open enough for easy auditing. Bitcoin cash has already a first implementation of the Schnorr signatures and is working hard on further improvements, with Bitcoin on its heels. There have been some BIP proposals to make this possible, with a new SegWit version 1 output type to allow spending rules based on Taproot, Schnorr signatures or MAST but also batch validation and signature hash improvements.[138] Therefore the new opcode OP_CHECKSIGADD has been proposed to allow for multisignature policies in a batch-verifiable way. This will probably be combined with some new OP_SUCCESS opcodes to allow the script to run more efficiently. You might be wondering why Bitcoin goes through such lengths to improve its privacy while there are already some other implementations possible and performed by other cryptocurrencies such as zcash (with zk-SNARKs) or Dash and Monero (ring signatures). Well, the main goal of these other cryptocurrencies is to provide privacy, while Bitcoin wants to remain an open network that wants to find a balance between privacy for its users and general adoption by the public. They want to focus on making the network more scalable and allow for transactions still to be audited, which is a requirement for several commerce applications and industries. This is why the developers have searched for and are developing these implementations so that there is a layer of privacy for the external world, while there is still the possibility for general adoption and acceptation by legislation.

2.17 BITCOIN MINING

We have gone in quite some detail when it comes to the Bitcoin network but we skipped one significant part: the mining hardware. Nowadays it is no longer profitable to start mining bitcoin from home with your pc or a GPU but this wasn't always the case. This is because of the mining difficulty that was introduced with the network. The goal is to mine a block every ten minutes, not slower nor faster. As more miners join the network, there is only one way to keep this speed and that is by increasing the difficulty and computer power needed to mine a block. In 2009, all you needed was your PC and the CPU to mine. You can still find articles from this period discussing the best processors for mining, aiming at $60 per CPU when you wanted to build a rig.[139] As more people started to join the network (we are 2011), it became too difficult to keep on mining by making use of the CPU and people started to switch to GPUs. It is used for complex computations and more specifically for those computers which have heavy graphics requirements. These units are much

[138] Wuille, P. (January 16, 2020). Bip taproot. *Github – Bitcoin bips*. https://github.com/sipa/bips/blob/bip-schnorr/bip-taproot.mediawiki. Accessed January 20, 2020.

[139] Edmonds, R. (March 8, 2018). Best CPUs for crypto mining. *Windows Central*. https://www.windowscentral.com/best-cpus-crypto-mining. Accessed December 18, 2019.

more powerful than CPUs (we are looking to an increase in power × 30). The next phase came with FPGA mining where the power was increased once more (FPGAs are between 3–100 times faster than the GPUs). It was 2013 when the ASIC miner joined the mining game and the entire level playing field changed. This is a piece of hardware that was specifically designed for the purpose of mining. Because of the continuous added competition, mining pools and farms came into existence. Mining pools allow you to join with your mining equipment with other miners to work together as one. The profits are shared based on the mining power you bring into the pool. Mining farms are companies that focus on building entire infrastructures, aiming to earn large amounts of bitcoins based on the power they have. Some concerns have been raised in the past because the majority of mining farms can be found in China.

2.18 BITCOIN RELAY NETWORKS

Closely linked to the concept of mining is the existence of Bitcoin relay networks. Again, this relates to the concept of time in the network. It is very important for miners to know when to start mining the next block when the new block is being propagated throughout the network. The relay network is used to minimize the latency in the network. The original network was introduced by Matt Corallo in 2015. It made use of specialized nodes on Amazon Web Services and connected the majority of the miners in the network. This implementation was replaced in 2016 with the introduction of "FIBRE" or "Fast Bitcoin Relay Engine" which was also created by Matt Corallo. It is an UDP-based network that relays blocks throughout the network and makes use of compact block optimization. Currently, developers are working on Falcon which is based on Cornell University research. It makes use of "cut-through routing" instead of "store-and-forward." With cut-through routing we enter a world of sequential routing where the messages are divided in units called "flits." These flits are of a very small size so that their header information must also be minimized, and this is done by forcing all these flits over the same path in sequence.

2.19 BITCOIN: THE CRYPTOCURRENCY

More famous than the network of course is the cryptocurrency behind the network, also called bitcoin. Bitcoin is being created by making use of the mining process and started at 50 bitcoin per block being created in January 2009. However, the rewards for mining are diminishing over time. Every four years[140] the rewards are being halved and now we have arrived at 12.5 bitcoin per block. Currently, over 85% of all bitcoins that will ever be mined, are already mined with a current reward of 12.5 bitcoins per block that will diminish to 6.25 bitcoins on the May 17, 2020.[141] If you follow this formula, we know that there will eventually come an end to the creation of new bitcoins. In

---

[140] Or more exactly, every 210,000 blocks.
[141] https://www.bitcoinblockhalf.com/.

the year 2140 there will be approximately 21 million bitcoin and after this no new bitcoins will be created ever again. Does this mean that there will be 21 million bitcoins on the market? No, not exactly, as bitcoins that are lost because of people losing access to their wallets can never be recovered. It has led to the famous saying that bitcoin is the equivalent of gold in the crypto world (with ether being the oil but more about that later). This mechanism makes sure that we are dealing with a deflationary currency (or at least it should be in theory). This means that the value of the currency will increase over time because we are dealing with a lowering supply while the demand increases over time. This makes sure that we have an increase in value, and the purchasing power increases. There are several ways that you can acquire bitcoin (if you are not a miner taking part in the mining process). First of all, you could make use of one of the several exchanges out there that offer bitcoin in return for other cryptocurrencies and/or regular currencies such as Euro or US Dollars. A second interesting approach is that of the "bitcoin ATM." These accept cash in return for bitcoin and send it to your (smartphone) bitcoin wallet. Finally, there is also the possibility to perform a direct trade with another person their wallet and/or perform services for bitcoin as you would any other currency.

2.20   PAYMENT CHANNELS ON BITCOIN

Over time several proposals have been made to create payment channels on top of the Bitcoin network (and other blockchain platforms). It is a technique that allows users to perform multiple transactions without committing all of the transactions to the Bitcoin blockchain.[142] Over the years several implementations have been proposed to introduce this feature to the network. Below I am going to give a short overview of several of these proposals. These examples clearly show how the network has evolved over time and how developers have tried (and succeeded) to tackle the scalability problem that often comes with blockchain networks. Later in the chapter, you will also find the lightening network explanation and implementation, which is arguably the most famous payment channel protocol to date.

2.20.1   NAKAMOTO HIGH-FREQUENCY TRANSACTIONS

This implementation made use of the nLockTime field. It was proposed by Satoshi Nakamoto and could be used in his view to contain payments of multiple parties where each of the participants could sign their own input. To agree on a new version, each participant must sign a higher sequence number, agreeing on the inputs and outputs of the previous state. There were also some other options to only agree to your output (SIGHASH_SINGLE), a pre-agreed default option could also be created with the nSequenceNumber and OP_CHECKMULTISIG but the problem was that

---

[142] Payment channels. *Bitcoin*. https://en.bitcoin.it/wiki/Payment_channels. Accessed October 8, 2019.

in the end, the design wasn't secure as a miner and a participant could work together and commit a non-final version of the transactions taking place, stealing from the other participants.

2.20.2 SPILLMAN-STYLE PAYMENT CHANNELS

Proposed by Jeremy Spillman on the Bitcoin-development mailing list and implemented in BitcoinJ, this technique creates a secure deposit combined with a second transaction by which the two parties can release the funds. This way the possible attack that could take place with Nakamoto high-frequency transactions, can't take place. These transactions work unidirectional as there is always a payer and a payee and it is not possible to reverse money back in the opposite direction. The payee needs to close the channel before a certain expiration time.[143] Problem here? The channel was vulnerable to transaction malleability (described in more detail below).

2.20.3 CLTV-STYLE PAYMENTS

Similar to Spillman-style payment channels, the CLTV-style payment channels are unidirectional payment channels that expire after a specific time. They became possible after the BIP-65 specification and the CLTV-soft fork that took place end of 2015. This channel is resistant against the malleability problem but still has limited use, as it only works in a unidirectional fashion. If we want to really scale the network, we need to be able to make payments in both directions, and if it were possible, with multiple parties.

2.20.4 POON-DRYJA PAYMENT CHANNELS

Poon-Dryja payment channels make use of funds that are locked in a two-of-two multisig. Commitment transactions of each party must be written and signed, before even the funding transaction is signed. Segregated witness is of key value here as it makes use of unsigned transactions and as such, it requires a transaction format that separates the transaction that is hashed for the txID and the signatures. It are bidirectional channels that have no expiration time and can be closed either unilaterally or bilaterally.

2.20.5 DECKER-WATTENHOFER DUPLEX PAYMENT CHANNELS

Christian Decker and Roger Wattenhofer introduced duplex payment channels in their paper.[144] It makes use of nSequence (the nSequence introduced by BIP-68) and consists out of two unidirec-

---

[143] Spilman, J. (April 20, 2019). Anti DoS for tx replacement. *Linux Foundation.* https://lists.linuxfoundation.org/pipermail/bitcoin-dev/2013-April/002433.html. Accessed October 8, 2019.

[144] Decker C. and Wattenhofer R. A. Fast and scalable payment network with bitcoin duplex micropayment channels. *Ethz.* https://tik-old.ee.ethz.ch/file/716b955c130e6c703fac336ea17b1670/duplex-micropayment-channels.pdf. Accessed October 13, 2019.

tional payment channels with indefinite lifetime. Between the funding transaction of the channel and the final transactions, there is something called an "invalidation tree," which contains the off-chain transactions that happen between the parties. The first version of the transaction has the longest relative lock time, while the next has a slightly smaller relative lock time, and so on. This channel can be closed by either party but it is most efficient if both parties work together to close the channel, as everything comes down to one single transaction on the Bitcoin blockchain. This channel can be extended to contain multiple parties.

2.20.6 DECKER-RUSSELL-OSUNTOKUN ELTOO CHANNELS

Christina Decker, Rusty Russell, and Olauluwa Osuntokun introduced the Eltoo payment channel in their paper on April 30, 2018.[145] This is one of the many ideas put forward by Blockstream and Lightning labs (as you will discover later in the chapter). The channel makes use of two transactions whenever there is an update taking place: the actual update transaction and a CSV-encumbered set-tlement transaction that spends the update transaction. To make this happen, this channel requires a new type of signing flag called SIGHASH_NOINPUT and OP_CHECKLOCKTIMEVER-IFY. The OP_CHECKLOCKTIMEVERIFY is not used to enforce any particular future time but is rather used to enforce an ordering of the update transactions so that each later update can spend an earlier one, but not vice versa. It does not require any punishment features as in Poon-Dryija but the main reason this channel is not being used yet is because of the SIGHASH_NOINPUT that still needs to be implemented. When this happens, this implementation will probably be used by the lightning network.

2.20.7 HASHED TIME-LOCKED CONTRACTS OR HTLCS

HTLCs are an integral part of the lightning network nowadays. It uses hashlocks and timelocks which lead to a situation where the receiver of a payment has to either acknowledge that he received a payment by generating a proof of payment or forfeit the payment altogether, after which it is returned to the payer. These hashed time-locked contracts can be combined with Poon-Dryja pay-ment channels which increases the security of the payments and without the necessity of recording these transactions on the Bitcoin blockchain.

2.20.8 TRANSACTION MALLEABILITY

Segregated witness was implemented to prevent all forms of transaction malleability. Before Segwit (and to this day for those nodes that didn't follow the soft fork required to implement segregated witness), there was work being done on researching all forms of transaction malleability. BIP-62

---

[145] Decker, C. and Russell, R. Eltoo: A simple Layer2 protocol for bitcoin. *Blockstream*. https://blockstream.com/eltoo.pdf. Accessed October 14, 2019.

was a work in progress to change the Bitcoin transactions so that malleability could be prevented.[146] Even though it is no longer being worked on, it is still an interesting study to see what forms of malleability were identified.

1. **Non-DER encoded ECDSA signatures:** older implementations (before v0.8.0 of the Bitcoin core) non-DER encoded signatures could still be relayed throughout the network.

2. **Non-push operations in scriptSig:** a sequence of operations in scriptSig, resulting in the intended data pushes, including more than only the push, results in a valid transaction.

3. **Push operations in scriptSig of non-standard size type:** there are several push opcodes in the Bitcoin scripting language, with each having different possibilities.

4. **Zero-padded number pushes:** number inputs in sciptPubKey opcodes can be zero padded.

5. **Inherent ECDSA signature malleability:** ECDSA signatures are malleable.

6. **Superfluous scriptSig operations:** extra data pushes at the start of the script which are not consumed by the corresponding scriptPubKey.

7. **Inputs ignored by scripts:** OP_DROP opcode can be used to ignore the last data push in a scriptSig.

8. **Sighash flags based masking:** these flags can be used to ignore certain parts of a script.

9. **New signatures by the sender:** the sender can create new signatures that spend the same inputs to the same outputs.

2.21   WASABI WALLET AND ZEROLINK

Privacy is always a concern when developers talk about the Bitcoin network. Zerolink is an implementation that focusses on fungibility of bitcoins and the privacy of the participants that perform transactions. The privacy is extended not to only a single transaction but also to chains of transactions. Zerolinks main goal is to break all links between separate sets of coins.[147] In the end, zerolink

[146]   Dashjr, L. (January 19, 2017) Bip-0062. Github – Bitcoin bips https://github.com/bitcoin/bips/blob/master/bip-0062.mediawiki. Accessed October 14, 2019.

[147]   Nopara73 (April 28, 2020) ZeroLink: The bitcoin fungibility framework. *Github – ZeroLink.* https://github.com/nopara73/ZeroLink?source=post_page. Accessed October 15, 2019.

is the combination of a wallet privacy network combined with Chaumain CoinJoin. CoinJoin was first introduced by Gregory Maxwell in 2013. It comes down to multiple participants adding inputs and outputs to a common transaction so that the transaction graph becomes obfuscated. The Chaumain CoinJoin uses chaum blind signatures. This means that each participant provides input and a blinded output to the Tumbler which signs the blinded output and gives it back to the participants. The participant then unblinds the output and provides it to the server in a signed form through a different anonymity network identity. The Tumbler eventually constructs CoinJoin transactions and requires the participants to sign. An implementation of this privacy-based Bitcoin wallet is called "Wasabi wallet" and is open for use.

2.22   META-COIN PLATFORMS ON TOP OF BITCOIN: COLORED COINS

There have been several implementations over time on top of the Bitcoin protocol. The first implementations that occurred focused on various techniques to add metadata to the existing Bitcoin processes. This was done by making use of unused transaction fields to encode extra information in the transactions. It was with the introduction of the OP_RETURN transaction script opcode that more possibilities arose to include more information directly on the blockchain. This also meant that a new type of coin saw the light of day: the so-called colored coins. [148]These colored coins are used to represent other cryptocurrencies but can also refer to digital and even physical assets. The first implementation of colored coins is called "Enhanced Padded-Order-Based Coloring," or "EPOBC," which assigns assets to one satoshi. Of course, the OP_RETURN opcode allows for more data to be stored, while these can also refer to external data which in turn can be used to reference specific assets. Over time there have been several implementations of which the most famous were OpenAssets (used by coinprism) and colored coins by Colu. However, nowadays the use of colored coins is mostly gone because of the introduction of the ERC-20 standard and others (explained later) which are easier and cheaper to use.

2.23   OPENASSETS PROTOCOL

Even though the protocol isn't used anymore that often (or not at all), it can still be interesting to play with the protocol. It can give you a better understanding of the workings of the Bitcoin network and how the developers wanted to interact with the network. How is the metadata association actually performed for assets? Well it works in three stages:[149]

---

[148]   These coins do not actually have a color, it is rather a reference to changing the attributes in some way, such as color would have done.

[149]   Charlon, F. (May 13, 2015) Open assets protocol. *Open Assets*. https://github.com/OpenAssets/open-assets-protocol/blob/master/asset-definition-protocol.mediawiki. Accessed October 17, 2019.

1. **Blockchain association:** the asset definition URL is embedded in the Bitcoin block-chain by the issuer;

2. **Asset definition file:** the definition file of the asset is made available at the URL stored in the blockchain; and

3. **Proof of authenticity:** SSL has to guarantee the identity of the issuer.

In the blockchain, the association can be performed by making use of an asset definition pointer that can have several formats but of which only one can only be used anymore because of limitation imposed in 2015 (marker output limited to 40 bytes) and to provide privacy for the parties exchanging the asset.

```
{
  "asset_ids": [
    "<base 58 asset id>"
  ],
  "name_short": "<string>",
  "name": "<string>",
  "contract_url": "<url>",
  "issuer": "<string>",
  "description": "<string>",
  "description_mime": "<mime type>",
  "type": "<string>",
  "divisibility": <integer>,
  "link_to_website": <boolean>,
  "icon_url": "<url>",
  "image_url": "<url>",
  "version": "<string>"
}
```

Figure 2.5: Asset definition.

The assets themselves, when included in transactions, give these transactions two new charac-teristics: an asset ID which is a hash of 160-bits, and the asset quantity being stored. The ID of an asset is the RIPEMD-160 hash of the SHA-256 hash of the output script referenced by the first input of the transaction that initially issued that asset. The transactions that make use of the pro-tocol, must have a "marker output." The marker output always starts with OP_RETURN and must have a PUSHDATA opcode containing a parsable marker payload.[150] The payload looks like this:

---

[150] If there are multiple payloads, only the first one is used and the rest is ignored.

Table 2.16: A parsable marker payload		
Field	**Description**	**Size**
OAP Marker	Always "0x4f41"	2 bytes
Version	The version number	2 bytes
Asset quantity count	Number of items in asset quantity list	1–9 bytes
Asset quantity list	0 or more LEB128-encoded unsigned integers	Variable
Metadata length	Length of the metadata field	1–9 bytes
Metadata	Empty or arbitrary data associated with transaction	Variable

There are two submodules where the first focuses on the transaction interpretation as colored coins while the second focuses on building the transactions themselves. If you would like to test out for yourself, do not hesitate to consult the Github page and make sure that you are deploying it on a Bitcoin core instance with both RPC enabled and the –txindex=1 parameter.

2.24 BITCOIN 2.0

Below you will find a substantial list of sidechains of the Bitcoin network but also new layers or protocol improvement projects. These all bring their own adaptations and innovations to the original Bitcoin network and try to offer a solution for an existing problem. Depending on what you are looking for, you will find some of these projects more interesting than others. Still it will show you that there is a major community supporting the Bitcoin network and looking for ways to improve the network so that there can be an even larger adoption of the network and the underlying technology. The Bitcoin network isn't dead; it is thriving and ever looking for new ways to ensure its future.

2.25 BITCOIN HIVEMIND

Hivemind is a P2P oracle protocol that wants to provide Bitcoin users with accurate data from outside the blockchain environment (see a more general description of oracles in the Ethereum chapter) so that participation in the prediction market (event derivatives) becomes a possibility. It is one of the implementations that sprung from Truthcoin (others are Amoveo, Augur, and Gnosis). The focus lies on the problem of information aggregation. The problem lies in the fact that we can never know when journalist, politicians, and so on actually tell us the truth. Similar to other projects such as Rootstock, they make use of merge-mining to ensure that the Bitcoin miners will also help to secure the Hivemind sidechain. To receive an extra incentive to do so, by receiving dividend revenue of the Bitcoin Hivemind. Within the network there would be the creation of two types of coins: CashCoins (CSH) and VoteCoins (VTC).[151] Participants with CashCoins can actually create so-called "prediction markets" of which the shares can be bought, sold or transferred and there is a value pegging in place where 1 CSH = 1 BTC. The VTC coins represent equity in the oracle corporations which are also called branches. Each of these branches has their own Vote-Coins. It is used to prevent Sybil attacks, provide proof of reputation within the network and can be used to punish participants that do not provide the necessary work within the overall network. The more VTC you have in a branch, the more voting power you have and the larger your share from the branch revenues. From the second you start to vote away from the majority or refuse to vote, you can lose your VTC coins in the branch. In a similar way you can gain more VTC coins if you participate in voting and with the majority. The distribution of VTC coins is based on a reputation-based redistribution scheme. This includes measures for deviant behavior, movements against the existence of an oligarchy and too volatile network environments. The system is also created in such a way that the minority vote gets the least reward (lower than 50% agreement on vote) and is maximized at 51% (preventing voting pools). These prediction markets can be created by any user which then becomes an "author." This author should have a certain belief that the market he is creating will be sufficiently liquid for trading and linked to this these authors only have an incentive to write "decisions" that by a certain day will be a locked fact. VoteCoin owners or "voters" have an incentive to maintain the long-run trading volume of their branch and have a strong incentive to vote on all decisions and vote the way they believe other voters will vote. The voters can be seen as the "employees" within the network. There are several concepts that you need to understand: vote, decision and ballot. The decision is what we are voting on. This can be something very simple as "Person A will be elected president." Or these can be much more complex statements on which the voters need to vote. Based on the decision, the voter will cast a vote (0 means false, 1 means true, NA means you didn't vote, and there are also some possibilities in-between when it comes to pricing calls for example). The ballot consists out of the matured decisions and the votes that have been cast on these decisions. At this moment the votes are revealed and the consensus algorithm determines the outcome of the vote. In case there wasn't a clear outcome of the vote, an "audit" takes place. Voters also have the right to vote after a waiting period and in case this is more than 50% of the blocks, there must be a re-vote.

The users of the CashCoins are the "customers" and don't have to interact at all with Vote-Coins if they don't wish to do so. They can trade on any prediction market they wish depending on what they think will increase in value. The trading is being facilitated by an automated market-maker as long as the voting is in progress. Between several rounds of voting there is a time called "Tau" or the intervote period. From the second the decision has been resolved, the trading ends and one enters the state of "redeeming." You can also understand that this is a very scalable

151 Sztorc, P. (December 14, 2015). *Truthcoin*. http://bitcoinhivemind.com/papers/truthcoin-whitepaper.pdf. Accessed November 18, 2019.

project in that sense that you can keep on creating more specialized branches. However, an unlimited amount of branches is undesirable, as it increases computational and economic cost, can create branches where voters no longer receive enough incentives to vote as they really want to vote and markets that are simply empty. That is why there is a system of "intelligent" splitting where voters have to decide whether or not they agree with a new branch. That is why there is a 'main' branch to start with, from which we can specifically start to split. If we refer to the image above, a market such as "football" could perfectly become its own branch, leaving a sports branch without football. If there is no activity left in branch for three consecutive intervote periods, it will be removed from the chain. The number of participants per branch will be limited to a number around 100,000 because of computational costs. This will mean that when the number increases, those with the lowest values will be removed from the branch. These participants would probably not have shared in any dividends or have participated in the process, leaving room for new active members to join. This limit only extends to the voters on the branch, not the owners! The current transaction speed for the Bitcoin network is too low for a competitive trading environment (even though low transaction speeds certainly help prevent double spending within the ordered list of transactions) and this is why Bitcoin Hivemind is looking at the GHOST protocol (or something similar) to provide a higher throughput speed.

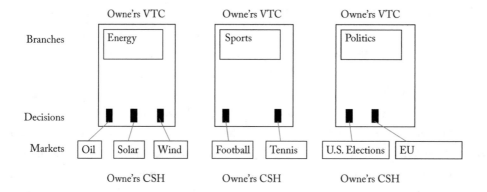

Figure 2.6: Bitcoin Hivemind.

2.26   THE MIMBLEWIMBLE PROTOCOL

For those of you that are Harry Potter fans, the phrase "mimblewimble" will be well known.[152] It is also the name of a protocol that was introduced in a paper on the July 19, 2016 by an author who

---

[152]   It is the curse that prevents one from speaking in the Harry Potter world.

called himself "Tom Elvis Jedusor" (oh yes, he did).[153] The author wanted to provide an answer to some of the privacy concerns that are linked with the Bitcoin network. The problem lies in the fact that all transactions must be stored in the blockchain while in fact they only relate to a small set of UTXOs. You cannot remove the other transactions because the final state and with it the UTXOs are only valid if the entire blockchain is valid. Transactions are also cryptographically atomic, which means the transaction graph can be used to identify users. The paper references several solutions put forward by Dr. Greg Maxwell, Nicolas van Saberhagen, and Shen Noether to improve the privacy within the blockchain but the author notes that these approaches lead to more data being used (KBs for confidential transactions), signatures that must be stored forever and the need for interactivity. The need for interactivity refers to the solution of Dr. Maxwell called "CoinJoin" where users can interactively combine transactions. CoinJoin is a trustless method of combining transactions by multiple spenders in one single transaction to make it more difficult for outsiders to determine who paid who. The number of transactions that are actually making use of this scheme is increasing over the years.[154] You do not need to change the Bitcoin protocol as such and the Wasabi wallet makes it easy to implement. This data storage requirement and need for interactivity was solved by Dr. Yuan Horas Mouton who made transactions freely mergeable but at a cost. Because he uses paring-based cryptography, everything becomes much slower. This solution was called OWAS or "one-way aggregate signatures." The proposal put forward is to remove Bitcoin script because it doesn't allow for merging of transactions with general scripts. Next, he took the idea of "confidential transactions" of Dr. Maxwell to allow authorization of spending outputs and the combinations of transactions without interaction (OWAS). This means that the Bitcoin structure and way of working is completely changed. Instead of individually signing each input and output there is a multisignature for all inputs and outputs. This is all thanks to the use of the Pedersen Commitment scheme where you no longer need addresses but instead use something called a "blinding factor" which is shared among the actors. Both inputs and outputs, together with the public and private addresses are encrypted so that only the involved parties know that they are involved in the transaction. So how does this Pedersen Commitment scheme actually work? First of all, the full nodes check whether the inputs and outputs totals are a balanced equation, making sure that no new coins are being produced out of thin air. The inputs and outputs totals are multiplied by a "blinding factor" that is made up out of both the private and public keys. This way the equation still has to hold but no one has an idea of the amounts that are involved while at the same time the involved parties can prove that they are the owners. The involved parties are eventually asked to sign a multisignature header to approve the transaction. But it also offers a solution for the storage of the data. While in fact there

[153] Jedusor, T.E. (July 19, 2016). MimbleWimble. *Scaling Bitcoin*. https://scalingbitcoin.org/papers/mimblewimble.txt. Accessed November 26, 2019.

[154] Manning, L. (May 1, 2019). Percentage of CoinJoin bitcoin transactions triples over past year. *Bitcoin Magazine*. https://bitcoinmagazine.com/articles/percentage-coinjoin-bitcoin-transactions-triples-over-past-year. Accessed November 6, 2019.

is a "cut through" feature which allows for much less information to be stored in the blockchain. In the Bitcoin world, each stage of an UTXO needs to be stored, as a proof of the transation of ownership throughout the transactions that took place. In the Mimblewimble world only the first input UTXO and the final output UTXO need to be stored, this way removing all the in-between steps and freeing up a lot of space. In January 2019, this Mimblewimble protocol was put into use with the launch of the Grin mainnet after years of development on the Github project started by "Ignotus Peverell" (Harry Potter fans rejoice).[155] There is also another active implementation called "Beam." Both implementations have their own specific features even though they make use of the same protocol. Grin has no cap on the number of coins that can be mined (60 per block) while Beam is limited to 263 million. Also, the proof of work consensus protocol differs: Grin uses the Cuckoo cycle (CuckARoo for 90% and CuckAToo for 10% while slowly shifting toward Cuck-AToo over time), and Beam equihash. However, the block time remains the same: for both it has landed on 1 min per block. What is the main difference with other coins that focus on privacy? You probably have heard of coins such as Monero, Dash, and Zcash which all bring their own ideas to the table when it comes to protecting the user from prying eyes. We will go in more detail when we discuss these coins separately but you should know that Zcash is much slower than either Grin or Beam, up to 64% of Monero inputs do not have the required "mixins" leading to traceable transactions and Dash is a centralized solution. Although we should note that these coins could try to incorporate the Mimblewimble protocol in their solutions, increasing the privacy even further.

2.27   THE ELEMENTS PROJECT

The elements project is another sidechain project created by Blockstream (it can also be run as a standalone project) that wishes to solve several problems that are encountered by participants in blockchain networks such as the lack of privacy, transaction latency and the risk to fungibility.[156] All these problems are being addresses by the developers behind the elements project by making use of federated block signing and confidential transactions. Block creation is no longer done by miners solving a proof of work algorithm but rather by "block signers," which are a federation of notaries who can create new blocks. In case Elements is being run as a sidechain, some notaries will take the role of "watchmen," making sure that there is a controlled and secure transfer between the two chains. Similarly to Mimblewimble, here the Elements project makes use of confidential transactions and a blinding factor. When using Elements as a sidechain, this is more often than not the Bitcoin blockchain and there will always be a two-way peg in place, allowing for the transfer of goods between the two chains. This ensures that Elements is interoperable with other block-chain platforms. It is the role of the watchmen to make sure that the token of the "main" chain

---

[155] https://github.com/mimblewimble/grin/blob/master/doc/intro.md.
[156] How elements works and the roles of network participants. *Blockstream*. https://elementsproject.org/how-it-works. Accessed July 13, 2019.

are effectively frozen and it is only when the transaction is verified that the equivalent amount of tokens on the Elements blockchain are released. The reverse transaction (from Elements to the main blockchain) is a little bit more complicated. First the "peg out" transaction is checked by watchmen who then sign the transaction that leaves a multi-signature wallet on the main chain. Only if enough members of the federation sign, the transaction becomes valid. In that case the tokens on the Elements chain are destroyed. Because of this way of working the speed of transactions and the creation of blocks is increased while the need for a 3rd party isn't necessary. When you make use of the Elements blockchain, you can create new assets all you like which is open to all network nodes (as long as you possess reissuance tokens which invoke the right of creation). Similarly, you can destroy tokens (if you possess these in your personal wallet). Furthermore, Elements is implementing new opcodes to the ones that already exist within the Bitcoin network (such as DETERMINISTICRANDOM and CHECKSIGFROMSTACK) allowing for more scripting possibilities and is similarly researching the possibilities of Schnorr signatures to further improve the efficiency of the network.

2.28   SIACOIN

Siacoin is another ambitious and interesting project, focusing on decentralized cloud storage. It is a decentralized way of storing files without a single point of failure. Because of the distributed way of working, there is no way to assure that enough nodes stay live and connected to the network for a participant to download his file. Well, they have a solution: Reed-Solomon erasure coding. What? This is how it works: when you upload a file to the Sia platform, your file gets divided in 30 segments which are distributed all over the world. This allows for maximum distribution and makes sure that at least a few of these nodes will stay online. Okay, now step two. Reed-Solomon codes were developed in 1960 by Irving S. Reed and Gustave Solomon. They proposed an encoding scheme which used a variable polynomial based on the target message where only a fixed set of evaluation points are known to both encoder and decoder. The decoder then has to generate potential polynomials based on subsets of unencoded message length of the encoded message length. This way the code can be used to detect mistakes that might have happened during transmission. Similarly, this technique is used by Sia. If 10 of the 30 divided segments of a file can be found, the entire file is available for download by the user. On top of this, each file segment is encrypted before it is distributed over the nodes by making use of the twofish algorithm. This way the users receive maximum protection from possible malicious participants. Finally, the up- and downloading facilities are facilitated by smart contracts so that no node can try to extort you for your file, it is locked in a secure way which does not require third party interventions. On top of that, the nodes have to pay collateral for each file contract, ensuring their participation and they receive "rent" from the participants that upload files. This rent is paid via micropayment channels. The host has to proof it is providing the storage in order to get paid, otherwise he is penalized. To increase adoption of the Sia network, there are several integrations with technology partners such as NextCloud (to host your own data storage similar to Dropbox), Duplicati (to allow for full computer backups) and Minio (distributed object storage server). To pay for all of this you can make use of Siacoin which is integrated with the blockchain solution of the platform.

2.29   THE COUNTERPARTY PROTOCOL

The counterparty protocol is another implementation on the Bitcoin network which wants to extend its use as much as possible. It is used to write specific digital agreements or even smart contracts with built-in scripts.[157] It allows for the creation of tokens for any type of asset digitalization or the creation of a cryptocurrency. To further enhance the speed, counterparty makes use of the lightening network (explained below) to speed up swaps between XCP and BTC. An even more interesting use case of the Counterparty protocol is the "asset exchange" where the roles of both escrow agent and clearing house are fulfilled by the protocol itself. The contracts exist on the network itself and funds from participating parties are immediately debited from their respective addresses. Only when the conditions of the contract is met, the distribution of funds can begin. Finally, they are also working on an alternative way of voting secured by their implementation of the Counterparty protocol and XCP-transactions. Securing both the identity of the participants and the result of the vote.

2.30   DROP ZONE

The original paper on Drop Zone was released in 2015 by "Miracle Max" but still receives attention on its ideas for creating a decentralized P2P market place.[158] It is another project that aims to work on top of the Bitcoin protocol.[159] The main idea (related to other P2P market places) is to prevent censorship and the freedom to sell and buy products. The difference here is that a blockchain is used as a solution. The project is not yet finished but it should work following these principles: a seller uploads the brief description of the good, an expiration and a hashtag. Buyers, on the other hand, search for goods close to their given location. In case the buyer finds a good it wants, the two parties can open a communication channel on the Bitcoin testnet so they can negotiate a price. Upon agreement, the buyer pays the seller on the mainnet and the location of the seller is revealed (GPS coordinates). However, the whitepaper also reveals a lot of weaknesses that need to be addressed. There is the vulnerability to Sybil attacks and reputation selling, but also non-fungible transactions,

---

[157] https://counterparty.io/platform/.

[158] (March 26, 2015). Drop Zone: P2P E-commerce paper. https://www.metzdowd.com/pipermail/cryptography/2015-March/025212.html. Accessed August 4, 2019.

[159] ScroogeMcDuckButWithBitcoin (2016). Drop Zone. https://github.com/17Q4MX2hmktmpuUKH-FuoRmS5MfB5XPbhod/dropzone_ruby. Accessed August 3, 2019.

abuse of API sources, unresolved sales, unscrupulous selling, and the fact that URL identifiers are centralized. It is still a project that has a lot of work to do but if established, might have a serious impact on P2P markets in an age where instances such as the SilkRoad have disappeared.

2.31   OMNI PROTOCOL

Mastercoin was the first implementation which tried to leverage the power of the Bitcoin network while at the same time building a new platform on top of it. The term "Master" was derived from "Metadata Archival by Standard Transaction Embedding Records." The main goal is to become a user-friendly platform where you can easily implement new development. It immediately also introduced the new cryptocurrency called, how could you even guess it, "Mastercoin" or "MSC." The idea put forward by the developers of Mastercoin is that their layer can be seen as the HTTP layer on top of the TCP/IP stack where the Bitcoin network becomes their protocol and Mastercoin the top layer.[160] This way they hope to open up development on the Bitcoin network to a much wider group of participants as you no longer have to be an "expert" to be able to participate and solve some common issues such as instability and insecurity. This was an attempt to stop the division of the developer community over all the rivalling alt coins. By pushing developers on top of the Bitcoin network and let them create their new cryptocurrencies on this protocol layer, the entire community would benefit as all developers would be pushing to a greater adoption of the network which would in turn lead to reward for every participant. The first release of their own coin was mainly intended to support their own growth so that developers could get paid to work on the new protocol layer. They even state they have an "exodus address," similar to the "genesis block" on the Bitcoin blockchain, which is the first Bitcoin address from where Mastercoins were generated.[161] It was later renamed to the "Omni protocol"[162] (and with it the OMNI token) and now aims at a broader target of developers including blockchain based crowdfunding where participants can send bitcoins or other tokens directly to the issuer and the sender receives the alternative token in return. The goal still remains to create a user-friendly developer environment where one can easily develop and release a cryptocurrency. First it made use of fake Bitcoin addresses and afterwards it made use of a multisignature scheme to embed data onto the Bitcoin blockchain. Currently, it makes in a large extent use of the OP_RETURN opcode of the Bitcoin network to lock data in the blockchain. Tether is probably the most (in)famous example of a cryptocurrency that makes use of the Omni protocol to leverage the power of the Bitcoin network. In the chapter on stablecoins you can find more information on this specific cryptocurrency.

---

[160]   https://en.bitcoinwiki.org/wiki/Mastercoin.
[161]   The Exodus address is: 1EXoDusjGwvnjZUyKkxZ4UHEf77z6A5S4P.
[162]   https://www.omnilayer.org/.

2.32   LIGHTNING NETWORK[163]

"We very, very much need such a system, but the way I understand your proposal, it does not seem to scale to the required size." – James A. Donald, November 2, 2008[164]

The above sentence may mean nothing to you but it was the first public comment made on the Bitcoin whitepaper that was put forward by Satoshi Nakamoto. Now, many years later, indeed scaling is one of the main issues that Bitcoin is dealing with. The system can only deal with about seven transactions per second, which is immediately the reason why they are so slow nowadays and why the transaction fees are so high.[165] Over the years the Bitcoin community has struggles with the limitations of the network and came up with several proposals. One of these is the lightning network. The general idea is that not every transaction needs to be recorded on the blockchain itself. It has added an extra layer on top of the Bitcoin blockchain so that any participant can create a channel with another participant to send and receive transactions nearly instantly with low or even no transaction fees. Several problems can be seen when we think about such payment channels. The paper written by Joseph Poon and Taddeus Dryvja largely focus on these. The first is the hostage taking problem. If you open a payment channel and you actually send the money, the other participant could simply say "I will never sign, unless you give me most of the transaction." To prevent this from happening, as you will learn from the description below, the money isn't actually sent when the payment channel is opened. This is because the transaction isn't signed until the very end. So both participants know what is going to happen, but it only happens when they both sign, so equally crossing the lines. Second, there is the problem of old commitment transactions.[166] The first transaction is the so-called "funding transaction" (to fund the payment channel) while all the other transactions happening between the participants are called "commitment transactions." If the payment channel is open for a long time, and we eventually want to close it, what prevents me from choosing an earlier transaction to close it on that is more in my favor? If this sounds confusing, think about this. Let's say you and I are funding the channel with both 1 BTC. A month ago you paid me 0.8 BTC, leaving the balances to 1.8 and 0.2 BTC. After 2 weeks I had to pay you 1.2 BTC, leaving the balances to 0.6 and 1.4 BTC. Why wouldn't I try to close on the balance leaving me with 1.8 BTC? Well, there are two securities built in. First, there is the opcode OP_CHECK-SEQUENCEVERIFY that uses the sequence field to freeze the output until enough miners have

[163]   Poon, J. and Dryja, T. (January 14, 2016). The Bitcoin lightening network: Scalable off chain instant payments. http://lightning.network/lightning-network-paper.pdf. Accessed October 21, 2019.

[164]   Donald, J.A. (November 2, 2008). Bitcoin P2P e-cash paper. https://www.metzdowd.com/pipermail/cryptography/2008-November/014814.html. Accessed August 9, 2019.

[165]   Visa has an average of 24,000 transactions per second with a peak capacity of over 50,000 transactions per second!

[166]   Bergmann, C. (April 29, 2017). The lightning network explained. Part I: how to build a payment channel. *Btcmanager.* https://btcmanager.com/lightning-network-primer-pt-i-building-payment-channels/?q=/lightning-network-primer-pt-i-building-payment-channels/. Accessed July 25, 2019.

confirmed that the output can actually be spent. This still doesn't prevent me from lying and taking the older transaction best of all it gives you time. Time to do what? Well, if I lie and close my part of the channel, you still have the option to take the whole fund of 2 BTC. What? Yes, as long as there are 2 participants watching the channel and waiting for the transaction to be accepted, all is fine. But if I lie, you have the nuclear option: you can just take it all. Now a bit more technical. The Lightning network works by making use of a multisignature wallet so that each participant can access this wallet by making use of their respective private keys, deposit some cryptocurrency and perform transactions between each of them. It is only when this wallet is closed and with it the channel, that the final results of all the transactions will be broadcast to the network, where it will be recorded as a single transaction. So what does this wallet actually look like? It makes use of something called a hashed timelock contract. It consists basically of a smart contract aimed at eliminating counterparty risk and implements time-bound transactions. It consists out of two participants where the first hashes his private key and sends it to the second participant. The first participant also generates a preimage which is used at the end to validate the transaction (a preimage is a datastring that is passed into a hash function and is sometimes called a "secret"). The second participant hashes his private key and sends it back to the first. Again, the second participant also generates a preimage by making use of a nominal transaction with the first participant. The first can now sign the transaction with the original key that is available in the preimage. The second participant can now do the same. There are still limitations though. The Lightning network does not solve the problem of the high transaction fees on the network. On top of that, the safest way of storing bitcoins is by making use of "cold storage." This is not possible as nodes making use of the Lightning network need to remain online at all times. Finally, it is still largely dependent on adoption of the network by the participants. As long it is not widely used, it does not live up to its full potential. Several parties developed an actual implementation on the lightning network. One of these is Lightning Charge by blockstream but an extensive list can be found on the following repository: https://github.com/bcongdon/awesome-lightning-network.

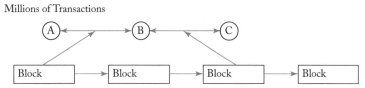

Figure 2.7: The bitcoin blockchain.

2.33   LIQUID NETWORK[167]

The liquid network is an implementation created by Blockstream, which in turn was founded in 2014 by a group of open-source coders behind Bitcoin. The network went officially live on September 27, 2018. It had an impressive range of launch participants: Altonomy, Atlantic Financial, Bitbank, Bitfinex, Bitmax, BitMEX, Bitso, BTCBOX, BTSE, Buull Exchange, DGroup, Coinone, Crypto Garage, GOPAX, Korbit, L2B Global, OKCoin, The Rock Trading, SIX Digital Exchange, Unocoin, Xapo, XBTO, and Zaif. The idea behind the liquid network was already started in 2015 with the white paper titled "Enabling blockchain innovations with pegged sidechains." Liquid is a federated sidechain which has been built on the Bitcoin network to facilitate faster bitcoin transactions between businesses and individuals, with a main focus on exchanges, financial institutions, and large traders. Small players can also use the network by making use of wallets or member exchanges. In the liquid network, you will not find the classic miners you would otherwise find but a group of so-called "block signers." These collect transactions into blocks, sign them and broadcast them over the network.[168] Within the network, block time has been reduced to a minute opposite to the ten minutes in Bitcoin. However, all the functionalities of the Bitcoin network are still present in liquid, so you can still create wallets, use block explorers and keys. As explained before, the liquid network is a sidechain, which means it has its own native cryptocurrency, called "liquid bitcoin" or LBTC which is backed with a two-way peg to bitcoin. The two currencies are interchangeable at any time. However, the network also leaves the possibility to issue other assets over the network that are linked to the real world. All of this information is safely encrypted so that it cannot be read by the public, while the hash of the asset can still be tracked through the transactions over the network. This way the transactions can still be audited and information can be shared with accountants and other interested parties.

2.34   ROOTSTOCK

So what is Rootstock? Rootstock is the (first) implementation that aims to build a smart contract platform (later explained in more detail) on top of the Bitcoin network. It aims to be the most secure decentralized network in the world (the Bitcoin network has the most hashing power to date in the network). Immediately, they also want to make their chain and smart contracts compatible with the Ethereum network, understanding and accepting their power and position in the market when it comes to these implementations and preventing a further divide in the community of developers supporting the different networks. They also have their own coin, called "smart bitcoin" or "SBTC," which is linked 1:1 with BTC (regular bitcoin). There is a two-way peg between RBTC and BTC so the coins are always interchangeable. Although the miners in the Bitcoin network are

---

[167] https://blockstream.com/liquid/.
[168] These block signers are often exchanges, market makers, and other big players.

also rewarded because of a feature which RSK calls "merge-mining." This means that the miners in the network will be effectively mining the RSK-chain and the Bitcoin chain, improving the security of both chains. The reward for the miners is 80% of the transaction fees while 17.5% goes to RSK labs, 1% to the RSK federation, 1% to RSK full nodes, and 0.5% to Bitcoin full nodes. The federation consists out of the set of semi-trusted notaries that need to sign the transactions described below. However, they can also provide oracle services (explained in Ethereum chapter) and other modules (check the docs for more information). To get a good-working relationship between their RSK-chain and the main Bitcoin network, they make use of a special hybrid relationship model. To lock the coins at the Bitcoin side of the network, they are using the classic drivechain implementation where the miners of the Bitcoin network need to vote and mine the block, effectively locking Bitcoin and unlocking SBTC. To reverse the transaction, the relationship is a bit more complex. To unlock the Bitcoin (and lock the SBTC), there is a combination of the multisignature approach combined with the drivechain approach. This means that both chains have to give their approval before this transaction actually can take place.

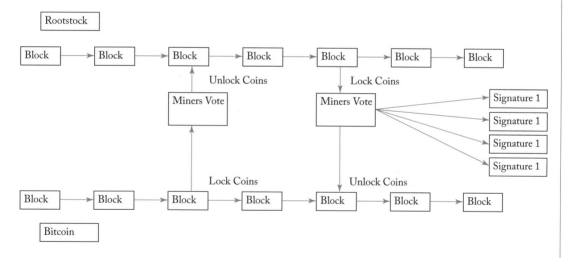

Figure 2.8: Rootstock Pegging System.

The ultimate goal is, when there are enough participants that engage in the relation between the two chains, this way of working will end. In the beginning, only the notaries can vote; then the combined relation will take place while the goal is to completely get rid of the signatures from the notaries and remove these completely so that only the drivechain relation remains. It is only the question if there will be enough participants so that this can actually be achieved. The implementation of smart contracts on Bitcoin has some consequences for other competitors in the market space such as Ethereum. This is another platform that brings the feature of smart contracts, and Ethereum is without a doubt the oldest and best-known brand when it comes to smart contracting and

DApp development. They wish to increase even more the adoption of the Bitcoin network but also allow industry to start developing applications on top of the Bitcoin network while still adhering to regulation. Their aim is to create (just like Ethereum) a Turing-complete chain with unlimited code size and memory with a persistent state and execution time. The consensus protocol in place is a combination of GHOST and DECOR with PBFT as a fallback mechanism which leads to a block time of 10 sec. While we will go into much more detail in the chapter on Ethereum, you just need to know for now that there is the possibility for smart contracts (and therefore DApps) on the RSK network. We also move away from UTXO scheme that is being used in Bitcoin to the use of accounts. The ECDSA algorithm is used to produce the public and private keys that are in use in the network. The smart contract in the Rootstock world consists out of several elements. The smart contract can both receive and send out messages combined with deposits or payments. Furthermore, the contracts have their own persistent memory and vaults to store a certain amount of SBTC. The next thing that Rootstock would like to achieve is to become a more scalable solution for Bitcoin than the lightening network. While the lightening network helps to increase the speed of the transactions, you still need a connection to the Bitcoin network and propagate the transactions through the network itself. Therefore, Rootstock is aiming at the "Lumino Transaction Compression Protocol," or LTCP, that is built on top of the RSK-network. This would lead to a transformation where Bitcoin can only do 3–5 transactions per second, to 100 transactions per second via RSK, to 2,000 on-chain transactions via the LTCP layer. On top of this layer, there should come the "Lumino" network that would be able to carry up to 20,000 of off-chain transactions. With this network, it would greatly overcome the limitations of the lightening network and the Lumino network would be able to reach billions of people (if you consider that each person settles their account once a month).

2.35   ZCASH

Zcash is one of the cryptocurrencies that in fact was built on top of the Bitcoin code base. The main goal of Zcash is once again to improve the privacy that is not offered by the Bitcoin protocol and to improve the transaction speed and scalability of the network.[169] On the other hand, the developers promise that the transactions can be auditable, but only with user permission. Within the world of Zcash, several transaction types exist as there are also two types of addresses: private or "z-addresses" and transparent or "t-addresses." As you might have guessed, a z-address starts with a "z" while a t-address starts with a "t." When a transaction takes place between two private addresses, the transaction is recorded but the addresses, transaction amount and the memo field are all encrypted to make sure that all this information remains private. A transaction between two public addresses remains completely visible when it becomes part of the blockchain. Tricky

---

[169]   https://z.cash/technology/.

are the transactions between a public and a private address. When a transaction takes place from a private to a public address we call this a "deshielding" transaction. The address and input remains encrypted but the receiving address and the amount are shown to the world. The reverse gets called "shielding" and also has the reverse implications. The developers also explain the implications of this increased level of security. If you send a balance, you normally need two addresses for the receiver: the receiving address and an address of your own to receive the remaining balance (unless you want to send to entire amount). This because sending a balance in the blockchain world actually means sending the entire balance that you have. If you just use your sending address, this is fine but not really private as you can create an identity profile based on this. One possible action might be to change the receiving address you use for the remaining amount to obfuscate anyone trying to uncover your identity. Not really helpful, as all transactions are linked and all these new addresses will just be linked one after another. Well hello, private (or shielded) addresses! I can just reuse my original sending address, as this is encrypted on the blockchain anyway.[170] But how does it provide this security? By making use of zero-knowledge proofs. These zero-knowledge proofs can be best explained by making use of the Ali Baba's Cave Parable put forward by Jean-Jacques QuisQuater (and others). In essence, the story goes that two people, A and B, stand in a ring-shaped cave with a door locking the end of the cave. B claims to know the secret to opening the door but does not wish to share this secret with A. Together they decide that A can choose which path B should take, and only if B comes out on the other side, it is proven that B actually knows the secret.

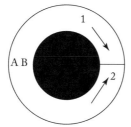 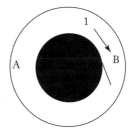

Figure 2.9: Ali Baba's cave.

2.35.1 ZK-SNARKS

Zcash makes use of something called "zk-SNARKs" or "Zero-Knowledge Succint Non-Interactive Argument of Knowledge" (so let's just keep saying zk-SNARKs). This is a form of cryptography where the prover can provide a proof of knowledge without actually revealing that knowledge AND without interaction between prover and verifier. This means that there is only a single message sent from prover to verifier. Currently, the most efficient way to achieve this is by making use of an initial

---

[170]   Peterson, P. (November 23, 2016) Anatomy of a zcash transaction. *Electric Coin.* https://electriccoin.co/blog/anatomy-of-zcash/. Accessed October 4, 2019.

setup where a common reference string, created with secret randomness, is shared between prover and verifier. If one could get access to the secret randomness to generate these parameters, one could attack the entire network and start creating counterfeit coins. Zcash went to extreme lengths to prevent this from happening. There was a multi-party ceremony with participants taking part so that the knowledge was well spread and afterward some of the computers were even blowtorched! To start, there is the creation of a public key–private key set with several participants in the multi-party ceremony. The reason behind this is that the private key linked to the public key needs to be destroyed. Otherwise, this would allow the participants to create counterfeit coins.[171] Because Zcash makes use of a multi-party process, all the participants their secret randomness is concatenated so that the private key will be destroyed unless all participants act maliciously. The final aspect we need to complete the process is the function that can actually determine whether a transaction is actually valid without revealing the secret information. Therefore, some of the Zcash network rules are encoded in the zk-SNARKs. "Succint" means that this proof is smaller and can be faster verified. Next, there is the "arguments" part of SNARKs which means that a malicious actor only has a very low probability of cheating the system because he has a limited amount of computing power. This might become a problem in the future of course with the rise of quantum computing. Finally, there is the "knowledge,'" which means that without the knowledge of the secret, the prover cannot create the proof. Now how does the algorithm actually work? The transaction validity function must be broken down to the lowest level, the arithmetic circuit where you can find single steps such as AND, OR, and NOT.

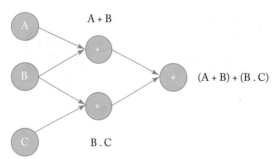

Figure 2.10: Arithmetic circuit.

Based on the outcome of these steps within the circuit, you come to what is called a "Rank 1 Constraint System" or "R1CS." This representation makes sure that all parameters have followed the correct path to the outcome. The verifier has a heavy computational task of checking the outcome

[171]   Gabizon, A. (September 25, 2016). Zcash parameters and how they will be generated. *Electric Coin.* https://electriccoin.co/blog/generating-zcash-parameters. Accessed November 11, 2019.

of each of these paths. With Quadratic Arithmetic Program (QAP)[172] this has been reduced to one computation of a polynomial. This in itself is again a great computational task but the verifying itself isn't that difficult. If an attacker were to generate a fake polynomial but doesn't know the right identity, the polynomial will fail almost every point so that only one point needs to be checked to verify the proof with a high probability. Below, we will give a short description of how the process actually functions and what we can find under the hood when we talk about zk-SNARKs. In essence, it comes down to "homomorphic hiding" (HH) functions.[173] They are similar to computational hiding commitments with that difference that hiding functions are deterministic functions of the input while commitments use additional randomness. For these HHs we have the following properties:

⇨ when $E(x)$, it is hard to find x;

⇨ for $x \neq y => E(x) \neq E(y)$; and

⇨ if we know $E(x)$ and $E(y)$, we can perform operations such as $E(x+y)$.

These expressions are used on a finite group Z^*_p instead of integers with the following properties:

⇨ there is a generator g so that all elements of Z^*_p can be written as g^a for some a in $\{0,\ldots,p\text{-}2\}$ (cyclic group);

⇨ with a large p, it is hard to find the integer a so that $g^a = h(\text{mod } p)$ (discrete logarithm problem); and

⇨ for a, b in $\{0, \ldots, p\text{-}2\}$ $g^a \cdot g^b = g^{a + b \, (\text{mod } p\text{-}1)}$.

From these properties one can know that $E(x+y) = g^{x+y \, \text{mod} \, (p-1)} = g^x \cdot g^y = E(x) \cdot E(y)$ or similarly that $E(ax+by) = (g^x)^a \cdot (g^y)^b = E(x)^a \cdot E(y)^b$.

Next, we have the polynomial P of degree d over a finite field F_p which is of the form $P(x) = a_0 + a_1 \cdot x + a_2 \cdot x^2 + \ldots + a_d \cdot x^d$ for $a_0, \ldots, a_2 \in F_p$. This polynomial can be evaluated for a point $s \in F_p$ where x is replaced by s.

Now we combine the principles of homomorphic hiding and the polynomial so that we can perform blind evaluation.[174] For two participants A and B, one knows the polynomial P while the other wants to learn $E(P(s))$ for a certain point s. However, the participants do not wish to share either P or s. This is solved by participant A sending $E(1), \ldots, E(s^d)$ to participant B who replies with $E(P(s))$. This way no one learns the true value from the other participant. The next logical step

---

[172] Gennaro, R., Gentry, C., Parno, B., and Raykova, M. (2012). Quadratic span programs and succinct NIZKs withpout PCPs. *IBM T.J. Watson Research Center*. https://eprint.iacr.org/2012/215.pdf. Accessed November 11, 2019.

[173] Gabizon, A. (February 28, 2017). Explaining SNARKs. *Electric Coin*. https://electriccoin.co/blog/snark-explain. Accessed December 3, 2019.

[174] Gabizon, A. (February 28, 2017). Explaining SNARKs. *Electric Coin*. https://electriccoin.co/blog/snark-explain2. Accessed December 3, 2019.

in the process is forcing participant B to actually reply with the truthful answer from his polynomial calculation. In this process we make us of the knowledge of coefficient test or "KC" test. The generator g we defined before becomes the generator of a group G with order $|G| = p$.[175] We also define $\alpha \in F^*_p$ where α is a pair (a, b) in G and $a, b \neq 0$ and $b = \alpha \cdot a$.

The KC test:

⇨ A computes $b = \alpha \cdot a$ and chooses at random $a \in G$,
⇨ A sends B (a,b),
⇨ B responds with (a', b') which is also an α-pair, and
⇨ A checks and accepts in case this is true.

The response by B can only be calculated by choosing a $\gamma \in F^*_p$ so that $(a', b') = (\gamma.a, \gamma.b)$ so that $b' = \gamma \cdot b = \gamma\alpha \cdot a = \alpha(\gamma \cdot a) = \alpha \cdot a'$ and therefore is an α-pair. The KCA or "Knowledge of Coefficient Assumption" states that this final statement is always the case. This sounds fine for 1 α-pair while there are multiple of these pairs being send out. It is up to the receiver to use these α-pairs together to create a new one following these properties:

$$c_1, \dots, c_d \in F_p \text{ and } (a',b') = (\textstyle\sum_{i=1}^d c_i a_i, \sum_{i=1}^d c_i b_i) => a' = \sum_{i=1}^d c_i \cdot a_i.$$

From which we can create the d-power KCA[176] in G where participant A chooses a set of α-pairs with a polynomial structure $\alpha \in F^*_p, s \in F_p$ and sends participant B the α-pairs $(g, \alpha \cdot g), (s \cdot g, \alpha s \cdot g), \dots, (s^d \cdot g, \alpha s^d \cdot g)$. Participant B knows $c0, \dots, c_d \in F_p$ (with a certain high probability) and then responds with $a' = \sum_{i=0}^d c_i s^i \cdot g$.

To reduce the calculations we have to do over the arithmetic circuit, we make use of QAPs (as we mentioned before). For a strongly reduced explanation below (find an excellent explanation by Ariel Gabizon on electriccoin.co), you can look at the following image.

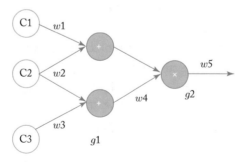

Figure 2-11: The arithmetic circuit.

[175] Gabizon, A. (February 28, 2017). Explaining SNARKs. *Electric Coin.* https://electriccoin.co/blog/snark-explain3. Accessed December 3, 2019.
[176] Groth, J. (October 26, 2010). Short pairing-based non-interactive zero-knowledge srguments. *University College London.* http://www0.cs.ucl.ac.uk/staff/J.Groth/ShortNIZK.pdf. Accessed October 1, 2019.

Participant B wants to proove he knows the inputs C1, C2, and C3 $\in F_p$ and that these lead to a certain outcome when combined. We make use of the arithmetic circuit to proof the outcome. In the image, the first set of wires are "the input wires"and the final wire "the output wire." Furthermore, if we would like to translate to a QAP the following properties hold:

⇨ if an outgoing wire goes to more than 1 gate, we still see it as 1 wire ($w2$),

⇨ multiplication gates have exactly 2 input wires (left and right), and

⇨ The addition gate and wires going from addition to multiplication gate aren't named.

The multiplication gates are associated with a field element g_x associated with a target point $\in F_p$ and we will also define a set of left wire polynomials (L_1, ..., L_d), right wire polynomials (R_1, ..., R_d) and outgoing wire polynomials (O_1, ..., O_d). These polynomials will usually be 0 on the target points, except for those involved in the target's point corresponding multiplication gate. [177]

When we start from fixed values c_1, ..., c_a => L := $\sum^a_{i=1} c_i . L_i$, R := $\sum^a_{i=1} c_i . R_i$ and O := $\sum^a_{i=1} c_i . O_i$. From this we can define the polynomial P := L . R − O.

We can state that c_1, ..., c_a is legal assignment if and only if P vanishes over all the target points. For this to work out there must be a target polynomial T that divides P. This is where QAP comes in as a QAP of degree d and size m consists of polynomials L_1, ..., L_d, R_1, ..., R_d, O_1, ..., O_d and a target polynomial T of degree d. The QAP can only hold if L = $\sum^a_{i=1} c_i . L_i$, R = $\sum^a_{i=1} c_i . R_i$, O = $\sum^a_{i=1} c_i . O_i$, P = L . R − O and T divides P. This all comes together in the Pinocchio protocol. The steps before only gets us to B providing the proof over the right degree of polynomial but doesn't force B yet to use the correct assignment c_1, ..., c_a. For each $i \in \{1, ..., a\}$ we create the polynomial $F_i = L_i + X^{d+1} . R_i + X^{2(d+1)} . O_i$ where we need to use X^{d+1} and $X^{2(d+1)}$ because we cannot mix the coefficient L, R or O in F. Therefore coefficient 1, X, ..., X^d are the coefficients of L, the next $d + 1$ coefficients X^{d+1}, ..., X^{2d+1} are the coefficients of R and the last $d + 1$ coefficients are those of O. With QAP we learn that F = $\sum^a_{i=1} c_i . F_i$, L = $\sum^a_{i=1} c_i . L_i$, R = $\sum^a_{i=1} c_i . R_i$ and O = $\sum^a_{i=1} c_i . O_i$ for the same coefficients c_1, ..., c_a and if F is a linear combination of the produced F_i's, we know that these were produced from the same coefficients. Again, we have to apply the homomorphic hiding functions when we want to know how zk-SNARKs function. Participant A sends a set of hidings $E(\beta . F_1(s))$, ..., $E(\beta . F_d(s))$ to participant B and asks B to produce the hiding outcome of the assignment $E(\beta . F(s))$. Problem is that participant A could try to deduce information of the hidden L, R, O, and T by applying a set of coefficients (c_1, ..., c_d), discovering whether the hidden L, R, T, and O he produces match or not and when this is not the case, he knows this is not participants B coefficients set. To prevent this, there is a random T-shift to each polynomial applied by participant B.[178] Now we need two more steps to complete the entire protocol: we need a

[177] Gabizon, A. (February 28, 2017). Explaining SNARKs. *Electric Coin.* https://electriccoin.co/blog/snark-explain5. Accessed December 3, 2019.

[178] Gabizon, A. (February 28, 2017). Explaining SNARKs. *Electric Coin.* https://electriccoin.co/blog/snark-explain6. Accessed December 3, 2019.

homomorphic hiding function that supports both addition and multiplication and we need to shift from an interactive system to a non-interactive system. For this to happen, we need to apply elliptic curves. You can find the explanation on elliptic curve cryptography in the chapter on "technologies to understand." This is combined with something called "Optimal Ate Pairing." By combining these two techniques, we arrive at a situation where we actually can compute the multiplication hidings based on the correct hidings of two other elements (an in-depth explanation would lead us too far from the description here). To allow for the non-interactive part, we make use of the common reference string (CRS). First, before any communication takes place between the parties, the CRS $(E_1(1), E_1(s), ..., E_1(s^d), (E1(1), E1(s), ..., E1(sd), E_2(\alpha), E_2(\alpha s), ..., E_2(\alpha s_d)) E2(\alpha), E2(\alpha s), ..., E2(\alpha sd))$ is published which is based on random data. These elements are hence used to help compute the a set (a, b) based on a random chosen $s \in F_p$ and $\alpha \in F^*_p$.[179] Verification follows by the other participant who verifies that $E(\alpha x) = \text{Tate}(E_1(x), E_2(\alpha)) E(\alpha x) = \text{Tate}(E1(x), E2(\alpha))$, and $E(y) = \text{Tate}(E_1(1), E_2(y))$ from $x, y \in F_r$ so that $a = E_1(x)$ and $b = E_2(y)$.

2.35.2　ZK-STARKS

Future implementations of cryptocurrencies wishing to preserve privacy might make use of a new implementation called "zk-STARKs," which refers to "Zero-Knowledge Scalable Transparent Arguments of Knowledge" and can act as a faster and cheaper alternative to zk-SNARKs. zk-STARKs remove the need for the initial trusted setup and also remove the computation difficulty (and the linked theoretical quantum computer attacks). Finally, the time needed to compute the outcomes of zk-SNARKs should be improved to further improve scalability of the network.[180] StarkWare industries is the first one to try and develop an industry ready implementation of this new way of working. In broad terms, STARK makes use of arithmetization. It is the reduction of the problem of verifying a computation to the problem of checking that a certain polynomial, which can be evaluated efficiently on the verifier's side, which is the succinctly part, is of low degree. The arithmetization used in zk-STARKs is composed of two steps: generating an execution trace and polynomial constraints, after which these are transformed in a single low-degree polynomial. The prover and verifier agree in advance on what the polynomial constraints are in advance so that, when the prover generates the execution trace, the prover can convince the verifier that the constraints have been satisfied without actually showing the execution trace. It makes use of Collatz conjecture to create a Collatz sequence execution tree which leads to the polynomial constraints. Starkware, together with 0x, has come up with StarkDEX which allows the creation of decentralized exchanges on top

[179] Gabizon, A. (February 28, 2017). Explaining SNARKs. *Electric Coin.* https://electriccoin.co/blog/snark-explain7. Accessed December 3, 2019.

[180] Ben-Sasson, E., Bentov, I., Hresh, Y., and Riabzev, M. (March 6, 2018). Scalable, transparent, and post-quantum secure computational integrity. *Israel Institute of Technology.* https://eprint.iacr.org/2018/046.pdf. Accessed November 23, 2019.

of the Ethereum blockchain with a higher throughput of trades compared to other current solutions and an improvement in the use of gas.

2.36   ZCASH AND HAWK

HAWK continues on the ideas of privacy and protection of personal information with a blockchain-based smart contract system that stores encrypted transaction on the blockchain itself. It is being promoted by students and faculty staff at Cornell University and the University of Maryland. The goal of HAWK is to allow non-technical people to create privacy preserving systems without having to know the details of cryptography.[181] The program is split in two parts: private and public. The private portion of HAWK is actually responsible for the encryption of the data while the public portion doesn't come into contact with the actual data.

2.37   HARD FORKS

This is only a limited overview of the hard forks that have happened or are planned. There were also so many in the past that failed to reach their target or simply never gained any support. Over the years there have been over 105 fork projects.[182] There was during the heights of the crypto-craze a whole set of attempts to fork Bitcoin which included: Super Bitcoin, Bitcoin Platinum, Bitcoin Uranium, Bitcoin Cash Plus, Bitcoin Silver, Bitcoin Diamond, and so on. Some crypto coins are already forks based on forks (such as Bitcoin stash which is a working fork of Bitcoin cash). I chose to list here some of the more significant changes that happened in the past while a hard/soft fork was still something considerable. Others in the future might still be considerable and might have significant impact on the network and the community. If you would like to go into detail in all the Bitcoin forks, be my guest. I am certain it will be an interesting journey although I am not sure where your journey will end. Each of the forks over time wanted to achieve something specific, to change something inherent to the Bitcoin protocol as it existed. Some of the forks focused on privacy, others on the transaction speed and still others on the block size and/or transaction costs. Another famous one developers like to focus on is the limited supply of Bitcoin. Either way, some of these were successful, or even more successful (up to your personal opinion) than the original Bitcoin release, others were doomed almost from the start.

[181]   Kosba, A., Miller, A., Shi, E., Wen, Z., and Papamanthou, C. (2015). Hawk: the blockchain model of cryptography and privacy-preserving smart contracts. *University of Maryland*. https://eprint.iacr.org/2015/675.pdf. Accessed November 23, 2019.

[182]   https://forkdrop.io/how-many-bitcoin-forks-are-there.

2.38   BITCOIN XT

One of the first (software) hard forks of the Bitcoin network was Bitcoin XT.[183] Its software was created by Mike Hearn in 2014 so that several new features could be included. It increased the number of transactions from 7 to 24 a second and increased the block size significantly to 8 Mb. The first proposal was released as BIP 64, so at first it was considered as a change proposal for the original Bitcoin network. BIP 64 called for a small P2P protocol extension that performs UTXO lookups given a set of outpoints. The first release of Bitcoin XT (version 0.10) included this change. Following the first release, some other changes were added to the protocol. Gavin Andresen published BIP 101 to call upon the increase in size of the Bitcoin blocks. This was implemented in Bitcoin XT on the August 6, 2015 but eventually reverted and made a place for the 2 MB-blocks that are also supported under the Bitcoin Classic protocol. While it had some initial success, Bitcoin XT was mostly abandoned by the end of 2015. Since August 2017, Bitcoin XT is a Bitcoin Cash client by default since release G.[184] For each of the Bitcoin Cash upgrades, there have been subsequent Bitcoin XT upgrades to keep on supporting the network (respectively called releases H and I).

2.39   BITCOIN CLASSIC

After the failure of Bitcoin XT, some of the community still supported some of the ideas behind this hard fork. Bitcoin classic proposed to increase the block size to only 2 MB instead of any bigger block size.[185] Very similar to Bitcoin XT, this was a software hard fork and was building on top of the Bitcoin Core reference client. The fork was less aggressive as Bitcoin XT and therefore was able to gain a level of support that remained for a certain time. After a while the control over the block size was put in the hands of the miners and the nodes.[186] Ultimately, on the November 10, 2017 Bitcoin Classic ceased operations after the New York agreement failed which also had as aim to increase the block size of the "Legacy Bitcoin chain" as it was called by Bitcoin Classic.[187] In the ending notes, you can clearly read that the developers call Bitcoin Cash "the last hope" for Bitcoin scalability and direct all their supporters in this direction.

2.40   BITCOIN UNLIMITED

The difference within this hard fork was that it would allow nodes and miners to decide for themselves what the size of their blocks would be and this without restarting the node or compiling

[183]   Reiff, N. (June 25, 2019). A history of Bitcoin hard forks. *Investopedia*. https://www.investopedia.com/tech/history-bitcoin-hard-forks/.

[184]   https://github.com/bitcoinxt/bitcoinxt/releases.

[185]   https://bitcoinclassic.com/devel/Blocksize.html.

[186]   Zander, T. (November 30, 2016). Classic is back. https://web.archive.org/web/20170202055402/https://zander.github.io/posts/Classic%20is%20Back/.

[187]   https://bitcoinclassic.com/news/closing.html.

new executables.[188] It was also the first implementation that allowed for Xthin (Xtreme thin blocks) where the inefficiency of receiving the same transaction twice or more is fixed. On top of that, the client allowed for parallel validation so that nodes can validate more than one block. It presents itself as not being a "hard fork" but rather a new form of consensus mechanism to allow change throughout the Bitcoin network based on a democratic voting system. Emergent consensus is introduced so that underlying incentive mechanisms might help consolidate a consensus on block size between minders, users, companies, wallets and holders. Nodes also have the possibility to set an excessive block size (EB) and acceptance depth (AD) to enable them to delay the acceptance of extra-large blocks from miners by orphaning their blocks until they have reached a certain depth in the blockchain. There was some interest in the implementation but eventually also this hard fork lost interest of the community. However, since version 1.1.0.0 the Bitcoin Unlimited client is compatible with the Bitcoin Cash client. There is still support for this hard fork and it has a steady community but time will tell if this currency will stand the test of time.

2.41   BITCOIN CASH

Bitcoin Cash or Bcash was the most successful hard fork of the Bitcoin network to date.[189] It was a response to the SegWit update because a part of the community refused to work with the updates of SegWit. They felt that certain changes, also introduced by BIP91 would favor those that saw Bitcoin as an investment mechanism rather than a means of payment. To this day, these differences can be seen in the setup of the Bitcoin network (which favor smaller blocks and see it mainly as a store of value) and the community supporting Bitcoin cash, which support it as a means of payment. The Bitcoin network was around this period (mid-2017) also dealing with rising transaction costs which further pressured the Bitcoin community. Eventually a group proposed a hard fork that would try to deal with all of these pain points. This protocol allows blocks of 8 MB, it comes with low fees and fast transaction times. They also promise reliable confirmations, which make the use of the coin much more scalable. More functionalities are on its way as there is a focus on faster block propagation, UTXO commitments, Schnorr signatures and more. The opcode OP_CHECK-DATASIG is already available allowing for oracles and advanced scripts within the network. There is still a certain competition with the original Bitcoin network to implement some of these features faster, to proof that one of them is the "true" future of Bitcoin in the long run. The hard fork was created in August 2017 and has achieved a huge market cap.

---

[188]   https://www.bitcoinunlimited.info/.
[189]   https://www.bitcoincash.org/.

2.42   BITCOIN SV

Within the camp of Bitcoin Cash there was a further division between two main camps. The first camp, which was supported by Roger Ver and Johan Wu, promoted a software entitled Bitcoin Adjustable Blocksize Cap (bitcoin ABC) which aimed at keeping the block size at 32 MB. A second camp led by Craig Steven Wright and Calvin Ayre wanted a software called Bitcoin Satoshi's Vision (Bitcoin SV) that would increase the block size to 128 MB.[190] They have a clear view on what they believe the vision of Satoshi Nakamoto was and aim to bring back certain aspects such as the opcode OP_RETURN. All of this to create a scalable solution that has a big block size. As always, only time will tell of their success.

2.43   BITCOIN GOLD

A second hard fork after the SegWit update (and following Bitcoin Cash) followed in October 2017. The main goal of the creators was to return to mining with GPU's against the current development of too specialized hardware. An interesting feature created by the developers was a so-called "post-mine" of 100,000 coins which were placed in an endowment to finance further development of this fork. Another difference with Bitcoin is its proof-of-work algorithm.[191] Several attacks have taken place on the coin with a major DDOS attack at the launch of the currency on the website, but later on (in May 2018) there was also a 51% hashing attack where 338.000 BTG were stolen with a value at the time of about 18 million U.S. dollars.[192] It still knows market success despite the setbacks it has had in the past.

2.44   BITCOIN DIAMOND

Bitcoin diamond launched in November 2017 and they promise faster transaction times at about 2–7 transactions per second being processed. On top of that, they offer lower transaction fees and want to encourage new users by lower prices of their cryptocoin.[193] There is, opposed to most other forks from Bitcoin, a higher cap on the total amount of cryptocurrency that can be mined. Instead of 21 million, the possible total amount has increased to 210 million. It also has a block size of 8 MB and offers connectivity to the lightening network. At the time of writing, the valuation is quite low but this could change over time depending on the changes that will be introduced in the future and, as always, the popularity of the currency itself.

---

[190] https://bitcoinsv.io/.
[191] https://bitcoingold.org/.
[192] Cimpanu, C. (September 4, 2018). Bitcoin gold delisted from major cryptocurrency exchange after refusing to pay hack damages. *Zdnet.* https://www.zdnet.com/article/bitcoin-gold-delisted-from-major-cryptocurrency-exchange-after-refusing-to-pay-hack-damages/. Accessed December 19, 2019.
[193] https://www.bitcoindiamond.org/.

2.45   BITCOIN INTEREST

In January 2018, a new interesting fork took place that rewarded participants based on mining and holding their tokens for a certain amount of time. When mining a block, the reward is split in one part for the miner (13.5 BCI) and one part for the interest pool (3.24 BCI). You can choose to participate in a weekly or monthly interest round and receive an even share (based on your input) of the generated interest. It is one of the few projects that managed to keep a "steady" price for the time being.[194] It has a very low valuation at the time of writing and it doesn't seem to change soon in the near future. It certainly shows how the clash between the different ideals behind Bitcoin (store of value vs a means of payment), can come to new and innovative ways of interpreting cryptocurrencies. It was also meant as a way of stabilizing the network so that market volatility could be contained as much as possible, creating a stable and trustworthy environment for everyone.

2.46   BITCOIN PRIVATE

In March 2018, another interesting implementation launched which was called "Bitcoin Private" and is based on both Bitcoin and Zclassic (which in turn is a fork of Zcash). It introduces both an alternative private way of transactions (with zero knowledge proofs) and at the same time uses a proof of work algorithm that is GPU resistant.[195] By combining these techniques, its aim is to be open for the community that wishes to participate in the mining, while at the same time helping to secure the privacy of all participants. Some interesting and user-friendly tools are offered to quickly test out the network and the possibilities of mining but currently there is not a very high market cap or valuation for this fork.

2.47   HOW TO FORK BITCOIN

We have now been talking a lot about the forks that have happened in the past on the Bitcoin network, so it seems only natural that we should discuss at least high level how one would be able to fork the network. Take into account that this is only a tutorial describing the process, while in reality you should think about possible attacks on your new network, glitches, scrutiny, and more. Hash power is of key importance, just as trust in what you are doing. As there are several ways of approaching this question, we will focus on the client software from Bitcoin Core. Reasons to try out the Bitcoin Core software, is because it is released under the MIT license (free of charge), there have been hundreds of developers working on it and it has because of this built-in features and security measures that could only help your own project a whole way along.[196] For the explanation

---

194   https://www.bitcoininterest.io/.
195   https://btcprivate.org/.
196   Baczuk, J. (May 24, 2019). How to fork bitcoin – Part1. *Medium.* https://medium.com/@jordan.baczuk/how-to-fork-bitcoin-part-1-397598ef7e66. Accessed September 19, 2019.

below, we based ourselves on the explanation provided by Jordan Baczuk. First of all, make sure that you have an environment ready to go and install all the dependencies. To be able to start a fork, you need to start from the original code of course which you can clone from https://github.com/bitcoin/bitcoin. Baczuk suggests to put an upstream remote to pull updates later, which is not the focus of this text but for a functioning coin in a live environment, this could be interesting to think about. Based on the latest version, we will branch of the latest release (here as dummy 1.0). Finally, we make sure that we can build the project and all dependencies are installed.

⇨ Sudo apt-get update

⇨ Sudo apt-get install software-properties-common libssl-dev libevent-dev libboost-system-dev libboost-filesystem-dev libboost-chrono-dev libboost-test-dev libboost-thread-dev –y

⇨ Sudo add-apt-repository ppa:bitcoin/bitcoin –y

⇨ Sudo apt-get update

⇨ Sudo apt-get install libdb4.8-dev libdb4.8++-dev -y

⇨ Git clone https://github.com/bitcoin/bitcoin && cd bitcoin

⇨ Git remote add upstream https://github.com/bitcoin/bitcoin

⇨ Git checkout v0.18.1

⇨ Git checkout –b 1.0

⇨ ./autogen.sh

⇨ ./configure

⇨ Sudo make

Now you are ready to go and start to make changes on the actual code such as renaming the project, change the address prefixes, work on message prefix bytes, RPC and P2P ports, seeds, max supply, distribution, block size, and more. Do not think this is a light exercise or that this will not require much effort from your part. An example is the renaming the project to a name of your choice, because in the code there are thousands of references to bitcoin, Bitcoin, or BITCOIN, which have to be replaced by the equivalents of your "new coin." Below you can find the proposed scripts with $NAME$ for the new name you would like to use. Be aware that the code below also changes all bitcoin urls, breaking the link to Bitcoin Core, so you will have to change these again if you want to make the documentation accurate.

⇨ Sudo apt-get install rename-y

⇨ Git clean –xdf

⇨ sudo find -type f -not -path "./.git/*" -exec sed -i 's/bitcoin/$name$/g' {} +

⇨ sudo find -type f -not -path "./.git/*" -exec sed -i 's/Bitcoin/$Name$/g' {} +

⇨ sudo find -type f -not -path "./.git/*" -exec sed -i 's/BITCOIN/$NAME$/g' {} +

⇨ sudo find . -iname "bitcoin*" -exec rename 's/bitcoin/$name$/' '{}' \;

⇨ sudo find . -iname "*bitcoin*" -exec rename 's/bitcoin/$name$/' '{}' \;

⇨ ./autogen.sh

⇨ ./configure

⇨ Sudo make

⇨ sudo find -type f -not -path "./.git/*" -exec sed -i 's/$Name$* Core developers/Bitcoin Core developers/g' {} +[197]

Congratulations! You just have completed the first step in creating your own coin based on the Bitcoin Core source code. Next, we can change the address prefixes to something more suitable for our new coin. An entire list of prefixes can be found online.[198] These values can be adapted in the source file src/chainparams.cpp.

```
base58Prefixes[PUBKEY_ADDRESS] = std::vector<unsigned char>(1,111);
base58Prefixes[SCRIPT_ADDRESS] = std::vector<unsigned char>(1,196);
base58Prefixes[SECRET_KEY] =     std::vector<unsigned char>(1,239);
base58Prefixes[EXT_PUBLIC_KEY] = {0x04, 0x35, 0x87, 0xCF};
base58Prefixes[EXT_SECRET_KEY] = {0x04, 0x35, 0x83, 0x94};

bech32_hrp = "bcrt";
```

These values you can adapt to reflect the values you would like from the prefix table. In the same file you can adapt the network parameters so that your new network will not conflict with the Bitcoin network when your coin is going live.

```
pchMessageStart[0] = 0xf9;
pchMessageStart[1] = 0xbe;
pchMessageStart[2] = 0xb4;
pchMessageStart[3] = 0xd9;
nDefaultPort = 8333;
nPruneAfterHeight = 100000;
m_assumed_blockchain_size = 280;
m_assumed_chain_state_size = 4;
```

You should always look out for values that are already in use by other networks or that are unlikely to occur in normal data. You can make use of several sources to make sure that your choices are valid and won't generate any conflicts.[199,200] After you implemented these changes, we should always make sure we run again:

---

[197] With this command you can fix licensing and copyrights but more links would have to be restored.

[198] https://en.bitcoin.it/wiki/List_of_address_prefixes.

[199] https://www.utf8-chartable.de/unicode-utf8-table.pl.

[200] http://www.asciitable.com/.

⇨ sudo make

To adapt the RPC and P2P ports, we should look in the src/chainparamsbase.cpp and src/chainparams.cpp files. With the first example below:

```
{
    if (chain == CBaseChainParams::MAIN)
        return MakeUnique<CBaseChainParams>("", 8332);
    else if (chain == CBaseChainParams::TESTNET)
        return MakeUnique<CBaseChainParams>("testnet3", 18332);
    else if (chain == CBaseChainParams::REGTEST)
        return MakeUnique<CBaseChainParams>("regtest", 18443);
    else
        throw std::runtime_error(strprintf("%s: Unknown chain %s.", __func__, chain));
}
```

And the second example:

```
99          pchMessageStart[2] = 0xb4;
100         pchMessageStart[3] = 0xd9;
101         nDefaultPort = 8333;
102         nPruneAfterHeight = 100000;
```

Also, these can be adapted to the values you want, depending on how you want the network to run (not interfering with frequently used ports, etc.). The seeds in the chainparams file are the nodes that new nodes will connect to first when they are synchronizing with the network. As you are starting a new network, it is better to comment these out (assuming you have no seed nodes).

```
116    vSeeds.emplace_back("seed.bitcoin.sipa.be"); // Pieter Wuille, only supports x1, x5, x9, and xd
117    vSeeds.emplace_back("dnsseed.bluematt.me"); // Matt Corallo, only supports x9
118    vSeeds.emplace_back("dnsseed.bitcoin.dashjr.org"); // Luke Dashjr
119    vSeeds.emplace_back("seed.bitcoinstats.com"); // Christian Decker, supports x1 - xf
120    vSeeds.emplace_back("seed.bitcoin.jonasschnelli.ch"); // Jonas Schnelli, only supports x1, x5, x9, and xd
121    vSeeds.emplace_back("seed.btc.petertodd.org"); // Peter Todd, only supports x1, x5, x9, and xd
122    vSeeds.emplace_back("seed.bitcoin.sprovoost.nl"); // Sjors Provoost
123    vSeeds.emplace_back("dnsseed.emzy.de"); // Stephan Oeste
```

For the coin distribution we have to look into the initial block subsidy (50 BTC for Bitcoin) and the block halving interval (210.000 blocks for Bitcoin), as there is no simple "max supply" parameter that we can adjust. In validation.cpp you can find the information you need to adjust. For the initial supply:

```
CAmount GetBlockSubsidy(int nHeight, const Consensus::Params& consensusParams)
{
    int halvings = nHeight / consensusParams.nSubsidyHalvingInterval;
    // Force block reward to zero when right shift is undefined.
    if (halvings >= 64)
        return 0;

    CAmount nSubsidy = 50 * COIN;
    // Subsidy is cut in half every 210,000 blocks which will occur approximately every 4 years.
    nSubsidy >>= halvings;
    return nSubsidy;
}
```

While the halving interval can be adjusted in the now well-known chainparams.cpp:

```
class CMainParams : public CChainParams {
public:
    CMainParams() {
        strNetworkID = "main";
        consensus.nSubsidyHalvingInterval = 210000;
        consensus.BIP16Exception = uint256S("0x00000000000002dc756eebf4f49723ed8d30cc28a5f108eb94b1ba88ac4f9c22");
        consensus.BIP34Height = 227931;
        consensus.BIP34Hash = uint256S("0x000000000000024b89b42a942fe0d9fea3bb44ab7bd1b19115dd6a759c0808b8");
        consensus.BIP65Height = 388381; // 000000000000000004c2b624ed5d7756c508d90fd0da2c7c679febfa6c4735f0
        consensus.BIP66Height = 363725; // 00000000000000000379eaa19dce8c9b722d46ae6a57c2f1a988119488b50931
        consensus.CSVHeight = 419328; // 000000000000000004a1b34462cb8aeebd5799177f7a29cf28f2d1961716b5b5
        consensus.SegwitHeight = 481824; // 0000000000000000001c8018d9cb3b742ef25114f27563e3fc4a1902167f9893
        consensus.MinBIP9WarningHeight = consensus.SegwitHeight + consensus.nMinerConfirmationWindow;
        consensus.powLimit = uint256S("00000000ffffffffffffffffffffffffffffffffffffffffffffffffffffffff");
        consensus.nPowTargetTimespan = 14 * 24 * 60 * 60; // two weeks
        consensus.nPowTargetSpacing = 10 * 60;
        consensus.fPowAllowMinDifficultyBlocks = false;
        consensus.fPowNoRetargeting = false;
        consensus.nRuleChangeActivationThreshold = 1916; // 95% of 2016
        consensus.nMinerConfirmationWindow = 2016; // nPowTargetTimespan / nPowTargetSpacing
        consensus.vDeployments[Consensus::DEPLOYMENT_TESTDUMMY].bit = 28;
        consensus.vDeployments[Consensus::DEPLOYMENT_TESTDUMMY].nStartTime = 1199145601; // January 1, 2008
        consensus.vDeployments[Consensus::DEPLOYMENT_TESTDUMMY].nTimeout = 1230767999; // December 31, 2008
```

In the same block above you can adjust the block time (PowTargetSpacing), the target difficulty (nPowTargetTimespan). Take into account that these are the parameters for the main net, while there are similar blocks for the test nets. The block size is even trickier. By now you have a little bit of an idea of the previous forks that took place and the "factions" that came into being because of these divisions. With the introduction of Segregated Witness within the Bitcoin network, we have to take into account the scriptSig data. The formula for the transaction weight so far is:

- transaction weight = base transaction * 3 + total transaction size;

- base transaction size = the size of the transaction serialized and witness data stripped; and

- total transaction size = transaction size in bytes serialized according to the description in BIP144 (including the base and witness data).

When you check the consensus.h file, you can see that the current weight is set at 4.000.000, so that the maximum block size is 1MB without witness data.

```
/** The maximum allowed size for a serialized block, in bytes (only for buffer size limits) */
static const unsigned int MAX_BLOCK_SERIALIZED_SIZE = 4000000;
/** The maximum allowed weight for a block, see BIP 141 (network rule) */
static const unsigned int MAX_BLOCK_WEIGHT = 4000000;
/** The maximum allowed number of signature check operations in a block (network rule) */
static const int64_t MAX_BLOCK_SIGOPS_COST = 80000;
/** Coinbase transaction outputs can only be spent after this number of new blocks (network rule) */
static const int COINBASE_MATURITY = 100;
```

Based on what you would like to include in your "new" cryptocurrency, you can activate certain BIPs. Some of these needed the approval of miners in the network to be activated. As you cannot perform these activations anymore, you can do it directly in the chainparams.cpp file. You can find a short example below, where you should adjust the values accordingly (remove or turn to 0).

```
consensus.BIP16Exception = uint256S("0x00000000000002dc756eebf4f49723ed8d30cc28a5f108eb94b1ba88ac4f9c22");
consensus.BIP34Height = 227931;
consensus.BIP34Hash = uint256S("0x000000000000024b89b42a942fe0d9fea3bb44ab7bd1b19115dd6a759c0808b8");
consensus.BIP65Height = 388381; // 000000000000000004c2b624ed5d7756c508d90fd0da2c7c679febfa6c4735f0
consensus.BIP66Height = 363725; // 00000000000000000379eaa19dce8c9b722d46ae6a57c2f1a988119488b50931
```

An important one to remember is to remove the checkpoint data that is hard coded in the application. We talk about the minimum work that should be in the chain (nMinimumChain-Work), assume by default that the signatures in ancestors of this block are valid.[201] With all the previous work, we arrive at the piece where most of you probably wanted to start: the genesis block. You can personalize the message you want to embed inside of your new currency, to what you believe in or would like to support with your new implementation. Remember that if you are serious, it is meant to last and that your message may just stand the test of time.

```
52   static CBlock CreateGenesisBlock(uint32_t nTime, uint32_t nNonce, uint32_t nBits, int32_t nVersion, const CAmount& genesisReward)
53   {
54       const char* pszTimestamp = "The Times 03/Jan/2009 Chancellor on brink of second bailout for banks";
55       const CScript genesisOutputScript = CScript() << ParseHex("04678afdb0fe5548271967f1a67130b7105cd6a828e03909a67962e0ea1f61deb649f6bc3f4c
56       return CreateGenesisBlock(pszTimestamp, genesisOutputScript, nTime, nNonce, nBits, nVersion, genesisReward);
57   }
```

---

[201] chainTxData is used to estimate the sync progress.

On top of that, you should also generate the coinbase transaction, public key that should receive the transaction, timestamp, and the nBits. To mine the block, you can use the cpp_miner program to start the process. To ease the mining of the genesis block, you can add an exception in the validation.cpp and pow.cpp files so that you can lower the difficulty of this first block. If you would like to be able to spend the first transaction, you will have to adapt the code a bit as this was not possible with the original Bitcoin Core setup. This change you would have to do in the validation.cpp file. With all these changes, you can adapt it further so that it reflects your vision of what a cryptocurrency should be. Of course, this example with Bitcoin, accounts also for all other cryptocurrencies that were originally based on Bitcoin. You might have to focus on other aspects as well, but as you should understand by now, you can adapt in the code what you want. Important is to keep in mind what you want to achieve and the security of the participants.

2.48   ALTCOINS BASED ON BITCOIN

It might shock you, or you might already know this, but a lot of the cryptocurrencies to this date make use in one form or another of the code base that was once provided by the Bitcoin network. The code is open source, so that it can be reviewed and adapted by developer communities that have their own idea of what a coin should be or what features are important to them. Have all of these "altcoins" been successful? Of course not. A lot of coins barely saw the light of day without any support. A community that underlies a coin is of upmost importance if it wishes to stay "alive." I am not going to list all cryptocurrencies that are based on Bitcoin but I will give you a couple of examples below of the most successful ones (to date of course). Previously, we had the hard forks on Bitcoin, which still heavily refer to Bitcoin and why they believe they are better or necessary. The following examples you will see here, have lost most of their references to Bitcoin and really want to stand on their own, with their own vision and specific issues they are trying to solve or focus on. Important to note is that, just as with the list of hard forks, this list is only a sample and a picture set in time. This means that when you are reading this book, there are probably already new and popular cryptocurrencies that aren't included here (just as there are already some popular cryptocurrencies that I haven't included). The reason is that I only make a specific description of what is out there, while trying to cover everything, would be an endless and even hopeless task. I hope those that I disappoint, can forgive me for this.

2.49   LITECOIN

The next cryptocurrency based on Bitcoin is called Litecoin, also sometimes called the silver if Bitcoin is gold. It promises transactions that are nearly instant and near-zero transaction costs.[202] The number of coins that can be mined in the Litecoin network is also limited, just as in the Bitcoin

---

[202] https://litecoin.org/.

network but the total number has increased to 84 million coins. An importance difference between the 2 networks is the block time: it has been reduced to 2.5 min. The mining algorithm in place is called "Scrypt" (which requires fewer resources than the mining algorithms in place for Bitcoin or Monero) and is just as in the case of Monero, designed to be more ASIC resistant so that CPUs and GPUs are the preferable method for mining Litecoin. Over time, Scrypt ASIC miners have come to the market but the majority of the miners in the network still seem to be participants that make use of CPU/GPU miners. The Scrypt algorithm is a proof of work algorithm that hashes the input value with a salt. So very similar to Bitcoin, Litecoin tries to provide a cryptocurrency but the focus is already much stronger on the practical use, as it offers a higher transaction rate.

2.50   DASH

Next in the list of cryptocurrencies that I would like to shortly introduce is called Dash, which was created on January 18, 2014 and originally named Darkcoin. Similar to Monero and Zcash, this currency aims to protect the privacy and anonymity of the participants in the network.[203] It forked the Bitcoin code and introduced except for privacy also quick transactions and the use of masternodes.

The total supply that will be generated by the network will amount to a maximum of 18 million coins of which the final coin will be mined around 2300. Still quite a way to go, don't you think? The mining algorithm in use is the X11-algorithm which was developed by Even Duffield.[204] It is another algorithm that wants to build in a resistance against ASICs. The name "X11" literally refers to the fact that 11 different hash functions have been incorporated into the algorithm: BLAKE, BLUE MIDNIGHT WISH, Grøstl, JH, Keccak, Skein, Luffa, CubeHash, SHAvite-3, SIMD, and ECHO. Very simply explained, a value is provided to the first hash function and the output of that function is provided to the next, and the next until the very end. You can clearly understand that the combination of all these hash functions made Dash more secure than Bitcoin was before but the goal, namely being ASIC resistant, is no longer achieved. We already mentioned shortly before the masternodes. These were installed to simplify the system in use within the Bitcoin and other networks. Masternodes are full nodes that have a bond of collateral (basically a stake in the network of 1,000 DASH), which "allows the users to pay for the services and earn a return on their investment." Another differentiator between Dash and other popular networks, is the split in block rewards. Every mining reward is split in 3 ways: 45% for the miner, 45% for the masternodes and 10% goes to the treasury. The treasury is used to fund the further development of the network and future Dash projects. It is the votes from the masternodes that determine the future development and projects of Dash. One of the main goals of Dash, is to really become a day to day currency,

---

[203] https://www.dash.org.
[204] Asolo, B. (October 30, 2018). X11 algorithm explained. *Mycryptopedia*. https://www.mycryptopedia.com/x11-algorithm-explained/. Accessed November 27, 2019.

both in the US but also for abroad, in countries that are in financial distress. So that people maintain a means of payment. It is investing in these countries but also research to further the future of blockchain as a whole.[205]

2.51   NAMECOIN

When you visit the website, you will learn that Namecoin is "an experimental open-source technology which improves decentralization, security, censorship resistance, privacy, and speed of certain components of the Internet infrastructure such as DNS and identities."[206] It came into being on April 18, 2011 thanks to a developer named "Vinced" following a discussion on Bitcoin and its possibilities. To make sure that miners wouldn't just simple economic incentives and jump on the token that would provide them with the most profit, merge mining was introduced. With merge mining it was possible to mine both Bitcoin and Namecoin and the same time. According to the website it has several possible uses, as it can enhance the protection of free speech (censorship resistance), attach identity information (GPG, OTR keys, etc.) to an identity of your choice, human-meaningful .onion domains, decentralized TLS certificate validation, websites with .bit top-level domain, and promote other technologies such as file signatures, voting, web of trust, notary services, and more. There are some clear similarities with Bitcoin, such as the proof of work algorithm in use and the maximum cap of 21 million coins. It also has a mining rate of about 10 min on average but the block size is a bit lower than that of Bitcoin. It is stated that about one-third of all Bitcoin miners merge-mine Namecoin. This gives the network a hashrate security that is far more than most other altcoins. Where we find the clear difference is in its ability to store data within the blockchain transaction database.[207] The market cap isn't that great, certainly not for a coin that was one of the earliest ventures to increase and enhance the possibilities offered by Bitcoin. Nevertheless, it is a coin and network worth mentioning and looking into as its ideas are to this day still innovative and worth considering toward the future. It has a loyal base of supporters that understand and push for the decentralized DNS system which could be the fundaments of internet privacy and censorship resistance. The coin makes also a clear reference to Aaron Swartz, who was a prominent internet activist.[208] He published a text with a proposal for so-called "Nakanames," which was one of the concepts that was later on implemented in Namecoin. [209]

[205] Sharma, R. (June 25, 2019). What is dash cryptocurrency? *Investopedia*. https://www.investopedia.com/tech/what-dash-cryptocurrency/. Accessed November 27, 2019.

[206] https://www.namecoin.org/.

[207] Frankenfield, J. (March 5, 2018). Namecoin. *Investopedia*. https://www.investopedia.com/terms/n/namecoin.asp. Accessed November 28, 2019.

[208] The smallest unit of Namecoin is called Swartz.

[209] Schwartz, A. (January 6, 2011). Squaring the triangle: secure, decentralized, human-readable names. https://web.archive.org/web/20170424134548/http://www.aaronsw.com/weblog/squarezooko. Accessed November 28, 2019.

2.52   DOGECOIN

Next on the list is a cryptocurrency that listens to the name "Dogecoin," referring to the well-known meme from the internet.[210] It first started out as a joke by Billy Markus in 2013, but quickly found a user base and now has a market cap of about 500 million dollar. Jackson Palmer was encouraged to help make the idea a reality and together they were able to eventually make Dogecoin a success.[211] It is actually based on Luckycoin, which in turn is based on Litecoin. Similarly to these coins, it makes use of Scrypt for its proof of work algorithm, making it ASIC resistant. However, the block time is only 1 minute for Dogecoin and there is no limit on the number of coins that can be produced. A remarkable occurrence in the history of Dogecoin, is the hack on the 25th of December 2013 of the Dogewallet platform, which lead to the theft of millions of coins.[212] To aid those that lost funds due to the hack, an initiative called "SaveDogemas" was set up to donate coins to those people that had coins stolen.[213] On January 2014, enough tokens were donated to refund all participants that were victim of the attack.[214] Later on the fundraising didn't stop, as the community behind Dogecoin funded the Jamaican Bobsled team to go to the Sochi Winter Olympics, to help build a well in the Tana river basin in Kenya and Nascar driver Josh Wise.

2.53   RAVENCOIN

Based on the Bitcoin code, some important changes were made during the development of Ravencoin (RVN). The maximum amount of coins was increased to 21 billion, the block reward times were reduced to a minute, and messaging capabilities were added. Where does the name come from? Fans of *Game of Thrones* will know that in the fictional world of Westeros, ravens are used as messengers who carry statements (messages) of truth, and in a similar way Ravencoin whishes to send the truth about who owns an asset.[215] The disadvantage of networks such as Bitcoin is that you cannot know if an asset is embedded in a bitcoin, therefore you can accidently destroy your asset. Also, using bitcoin to embed the asset has a cost as well. Ethereum offers a solution but several ERC20 tokens can carry the same name, which in turn could again lead to the wrong asset being

[210]   https://knowyourmeme.com/memes/doge.

[211]   https://dogecoin.com/.

[212]   The platform's filesystem was hacked and the send/receive page was modified in such a way that all transactions were sent to a specific static address.

[213]   Couts, A. (December 27, 2013). Such generosity! After Dogewallet heist, Dogecoin community aims to reimburse victims. *Digital Trends*. https://www.digitaltrends.com/cool-tech/dogecoin-dogewallet-hack-save-dogemas/. Accessed November 29, 2019.

[214]   Feinberg, A. (December 26, 2013). Millions of meme-based Dogecoins stolen on Christmas day. *Gizmodo*. https://gizmodo.com/millions-of-meme-based-dogecoins-stolen-on-christmas-da-1489819762. Accessed November 30, 2019.

[215]   Tolu. (April 8, 2019) Important facts about the Game of Thrones-Inspired Ravencoin (RVN) *Bitcoin Exchange Guide*. https://bitcoinexchangeguide.com/important-facts-about-the-game-of-thrones-inspired-ravencoin-rvn/. Accessed July 3, 2020.

transferred or even destroyed. Ravencoin offers a solution by creating a network that is asset-aware and where you can make use of messages and voting. To make it ASIC-resistant, the network makes makes use of an algorithm called ×16r, where in fact 16 mining algorithms are constantly changed in order.

2.54 PEERCOIN

Peercoin calls itself "the pioneer of proof of stake" and is based on a paper that was released on August 2012 by Scot Nadal and Sunny King (who is also the creator of Primecoin). As you might have guessed, it is a proof of stake-based network that generates new coins based on the holdings of individuals. The network does still contain a proof of work component, making it more a hybrid system than a pure "proof-of-stake" network. The idea is that the proof of work algorithm becomes increasingly difficult, so that participants are more and more rewarded via a proof of stake system, eventually phasing out the proof of work aspect of the network. Peercoin also clearly sees cryptocurrency more as a store of value, as your chances of reward increase over the time you are actually holding the coins in your wallet. The main goal when this network was created, was to reduce the high-energy consuming proof of work algorithm that is in use by the Bitcoin network, but also wanted to provide increased security and energy efficiency.[216] There is also no limit on the number of peercoins that can be generated, as the proof of stake algorithm ensures a 1% yearly inflation of the minted coins. Next to that, the block time was around the starting time about 7 min (opposed to the 10 min in the Bitcoin network) and the transaction fees within the Peercoin network are driven by the protocol itself.

2.55 GRIDCOIN

Gridcoin has a unique approach to blockchain technology as it focuses on crowdsourcing of calculations for the scientific community.[217] It was published on October 16, 2013 by Rob Halförd and was constructed in such a way that proof of research is applied. Participants in the network are rewarded based on their computational contribution to science on BOINC (Berkely Open Infrastructure for Network Computing). Similar to Peercoin, Gridcoin also makes use of a proof of stake validation scheme so that it may become more ecofriendly than the Bitcoin network. Participants receive a reward of 1.5% on the staked coins while they can receive a payment on top based on the participation of the user on BOINC projects that are whitelisted. A lot of guides and examples are

---

[216] Frankenfield, J. (July 5, 2018). Peercoin. *Investopedia*. https://www.investopedia.com/terms/p/peercoin.asp. Accessed November 30, 2019.
[217] https://en.bitcoinwiki.org/wiki/GridCoin.

provided on the website to show how you might contribute and be rewarded as you are going solo, as part of a pool or rather just invest in the token.[218]

2.56   PRIMECOIN

The last token I would like to describe here is called "Primecoin."[219] It was launched on July 7, 2013 by Sunny King. Blocks are generated every minute and block difficulty is changed every block and the block reward is adjusted based on the difficulty of the block. What makes this network unique is the proof of work algorithm in place. It is based on the computation of prime numbers based on Cunningham chains and bi-twin chains. The results of the computation are stored on the block-chain, and shared with the scientific community for further research. The symbol of the network is the Greek letter psi, referring to Riemann and his zeta function. You can actually check in each block the primes by making use of the "getblock" output and combine it with the "primeorigin" field.

[218]   https://gridcoin.us/.
[219]   http://primecoin.io/.

CHAPTER   3

Ethereum

If crypto succeeds, it's not because it empowers better people. It's because it empowers better institutions. - Vitalik Buterin

Ethereum, or "the world computer" as it is often called by the supporting community, is a deterministic but practically unbounded state machine, consisting of a global singleton state and a virtual machine that applies changes to that state.[220] This seems to be a very complex sentence but it shows really well what the Ethereum network actually is. In the following sections we will start to dissect every part of the sentence stated above and try to explain what makes it so very different from Bitcoin, and why it is also the first proponent of a new type of blockchain, introducing concepts of its own, and more importantly: blockchain 2.0. The reason why this is important is that Ethereum was developed to fix an existing problem, the problem that the Bitcoin network wasn't immediately able to solve. Vitalik Buterin, the inventor of Ethereum, was a Bitcoin enthusiast and believed in the possibilities of blockchain but encountered a rather difficult problem when it came to development on the platform. As we have described before, the possibilities for development on the Bitcoin platform is rather difficult and limited. Combined with the fact that possibilities are more and more limited due to the fact that certain scripting modules are being removed to increase the stability and security of the Bitcoin network, creating applications on top of this network is quite difficult. There was of course the possibility to create a layer on top of the network but this removed, in part, the possibilities and opportunities that were brought by the blockchain technology itself. Knowing that this could be done differently, Vitalik Buterin published his white paper on Ethereum[221] in 2013 where he defended his idea of what could unlock the full potential of blockchain technology by creating a universal state machine. His efforts were soon supported by a lot of other people willing to work on this new solution. The network also came with some clear goals toward the future participants and developers that would join the network. First of all, it makes use of something that is called "the sandwich complexity model" where the goal is that the bottom-level architecture of the network should be as simple as possible. This means that the Ethereum platform wants to be as user-friendly as possible and that all things concerning network protocol, machine-level language, serialization, and so on shouldn't be that developers' main concern. This is, whenever possible, pushed to the other layers so that the developer can focus completely on the development part of

---

[220]   Antonopoulos, A. and Wood, G. (2018). *Mastering Ethereum*. 1st ed. California: O'Reilly Media.

[221]   http://blockchainlab.com/pdf/Ethereum_white_paper-a_next_generation_smart_contract_and_decentralized_application_platform-vitalik-buterin.pdf.

smart contracts and dApps in high-level languages. Secondly, the network focuses on freedom and generalization. This means that the developers behind the Ethereum platform believe in the values of "net neutrality" and do not wish to judge over which applications and smart contracts are developed on top of the network. They also aim to make the possibilities as broad as possible for those people that wish to develop on top of the platform. This means that there are opcodes defined for which currently there doesn't seem a clear use but this can always change in the future. This generalization goes quite far because the developers of the network refuse to build in even very common features. The reason is that they want to keep the protocols and the network open for everyone and if you want a specific feature to be used, you can perfectly define it yourself in a smart contract. The responsibility is up to you and you alone. Finally, there is no risk aversion, which means that the choices of the network include faster block times, generalized state transitions, and so on. This is again in part with the user and developer in mind so that they can have the best experience while maintaining some general principles.

3.1　THE ETHEREUM VIRTUAL MACHINE

An important concept to understand in Ethereum is the "Ethereum Virtual Machine" or EVM for short. It was first defined by Gavin Wood in his yellow paper on improvements for the Ethereum network.[222] In short, the EVM is a 256-bit register stack which is designed to run the code presented by a smart contract. It creates a level of abstraction between the executing code and the executing machine. This allows for a separation both between applications as well as their hosts. This is one of the key features that we want to see in smart contracts, as you will later see in the section called "smart contracts." These smart contracts are written in a programming language that is called Solidity (even though Vyper or Bamboo are also viable options).[223] These separate languages came into being because in the EVM a separate language must be used to code the applications. The EVM should be seen as a generalized, secure, ownerless, virtual machine.[224] This virtual machine will accept the programs running on top which we know as smart contracts. These will always produce the same result given a specific input (which means they are "deterministic"), which implies that the underlying state changes will also be the same. Every possible task that can be executed by a computer can in some way also be executed by the EVM, making it Turing complete. However, there is one limitation: the EVM is bound by the use of gas that must pay for the computations that are taking place. These smart contracts cannot be executed directly by the EVM but need to be compiled in a set of low-level machine instructions. These instructions, or "opcodes" allow the EVM to be Turing-complete. This stack-based way of working can remind one of how

---

[222] You can always check the yellow paper called "Ethereum: A Secure Decentralized Generalized Transaction Ledger" by Gavin Wood (2018).

[223] Deprecated programming languages are Serpent and Mutan.

[224] Dannen, C. (2017). *Introducing Ethereum and Solidity*. 1st ed. New York: Apress.

Bitcoin scripting works and in part this is certainly true, only the possibilities with Ethereum are much, much broader. Each of these opcodes is 1 byte so there can only be 256 opcodes (for now we already have 140 unique opcodes) which can be divided in a specific set of categories:[225]

- stack-manipulating codes,

- memory-manipulating opcodes,

- storage-manipulating opcodes,

- environmental opcodes,

- program counter related opcodes,

- halting opcodes, and

- arithmetic/comparison/bitwise opcodes.

A sample set of these opcodes can be found in Table 3.1.

Table 3.1: Sample set of opcodes

Opcode	Name	Description	Gas
0x00	STOP	Halts execution	0
0x01	ADD	Addition operation	3
0x02	MUL	Multiplication operation	5
0x03	SUB	Subtraction operation	3
0x04	DIV	Integer division operation	5
0x05	SDIV	Signed integer division operation (truncated)	5
0x06	MOD	Modulo remainder operation	5

Because the goal of the Ethereum network is to work efficiently, these opcodes are stored in so-called bytecodes. During the execution of the code, the bytecode is being split in bytes, each representing an opcode. However, there are also limitations to the use of the EVM as it uses a 256-bit register stack from which only the 16 most recent items can be manipulated at once. Furthermore, the stack can only hold 1024 items. This is why the opcodes use contract memory to retrieve data. To store this data indefinitely, one has to make use of storage. However, writing to storage is very expensive! The EVM is just as any other computer: always looking for changes to its state. Every time it encounters instructions of any kind, it will start to translate and run its own code. Each change the machine makes in its state will be based on the previous state. This makes sure that these changes are linked to each other and that changes in memory are made for a reason. The EVM is in a non-stop loop checking for new instructions to be executed. What is specific to Ethereum and

[225] Do not hesitate to consult the reference with all the opcodes on: https://ethervm.io/.

differentiates it from other computers is that the network here will constantly check for transactions taking place and the state of the EVM is the representation of the balances that at that moment of time exist. Of course, this brings with it a whole set of problems of its own.

> *"There exist far more invalid state changes than valid state changes. Invalid state changes might, e.g., be things such as reducing an account balance without an equal and opposite increase elsewhere. A valid transition is one which comes about through a transaction."*
>
> – Gavin Wood, Yellow paper

The ultimate goal of the EVM is to create a representative and trustworthy history that reflects each legitimate change in its state.

3.2   NETWORK COMMUNICATION IN ETHEREUM

The Ethereum network makes use of the RLPx transport protocol—a TCP-based transport protocol and named after the serialization format used by the network—to allow communication between the nodes that make up the network.[226] The protocol makes use of "Elliptic Curve Integrated Encryption Scheme" (ECIES') and is an asymmetric encryption method. ECIES in RLPx consists of the following parts:

- Secp256k1 elliptic curve with generator G,

- the NIST SP 800-56 Concatenation Key Derivation function,

- HMAC using the SHA256 function, and

- the AES128 encryption function—CTR mode.

All the cryptographic operations are based on secp256k1 and the nodes are required to keep a static secp256k1 static key. However, before there can be any communication there is another step that needs to happen. The initial connection is created by making use of a handshake between the nodes. There is first an "auth"-message send by the initiator that wishes to make a connection. The receiver decrypts the message and verifies if the message has been correctly signed (signature = keccak256(ephemeral-pubk)). The receiver will also derive the secrets from the message. Only if this is the case, it responds with the "auth-ack"-message to the initiator together with the first encrypted frame containing the "Hello"-message. The initiator derives in turn the secrets from the auth-ack and replies in kind with the first frame containing the "hello"-message. Both will authenticate this first frame and the handshake is complete if the MAC of the first encrypted frame is valid on both sides.

The hello-message looks a bit like this:

---

[226] https://github.com/ethereum/devp2p/blob/master/rlpx.md.

(example taken from https://github.com/ethereum/devp2p/blob/master/rlpx.md):

```
[protocolVersion: P, clientId: B, capabilities, listenPort: P, nodeKey: B_64, ...]
```

Once the authentication is complete, the communication between the nodes can begin. Important is that all messages after the authentication are framed. This allows multiplexing multiple capabilities over a single connection and because of the demarcation points that are created for message authentication codes, encrypted communication is easier. The multiplexing allows messages with different capabilities to be transferred over the same connection. Furthermore, the message authentication in RLPx makes use of 2 keccak256 states: egress-mac (sent) and ingress mac (received).

3.3 BLOCKS AND CHAINS

⑦ Block Height:	**0** ‹ ›
⑦ Timestamp:	ⓘ 1435 days 18 hrs ago (Jul-30-2015 03:26:13 PM +UTC)
⑦ Transactions:	8893 transactions and 0 contract internal transaction in this block
⑦ Mined by:	0x0000000000000000000000000000000000000000 in 15 secs
⑦ Block Reward:	5 Ether
⑦ Uncles Reward:	0
⑦ Difficulty:	17,179,869,184
⑦ Total Difficulty:	17,179,869,184
⑦ Size:	540 bytes
⑦ Gas Used:	0 (0.00%)
⑦ Gas Limit:	5,000
⑦ Extra Data:	◆◆◆N4{N◆◆|◆p◆◆3◆◆◆i◆◆z8◆◆ ◆◆ (Hex:0x11bbe8db4e347b4e8c937c1c8370e4b5ed33adb3db69cbdb7a38e1e50b1b82fa)
⑦ Hash:	0xd4e56740f876aef8c010b86a40d5f56745a118d0906a34e69aec8c0db1cb8fa3
⑦ Parent Hash:	0x0000000000000000000000000000000000000000000000000000000000000000
⑦ Sha3Uncles:	0x1dcc4de8dec75d7aab85b567b6ccd41ad312451b948a7413f0a142fd40d49347
⑦ Nonce:	0x0000000000000042

We already went through a lot of information regarding the Ethereum network and how it is developing over time. As before, we will start from the top and move down to how the Ethereum blockchain (currently) looks like. We will go through the several parameters that make up blocks in the blockchain and it is always interesting to start at the very beginning: the genesis block.

The block height is obviously 0 as it is the first block and the timestamp gives us July 30, 2015. There are a lot of transactions included which refer to the first addresses receiving the first reward from the genesis block. Below you can see a short example from what these transactions look like.

Txn Hash	Block	Age	From		To	Value
GENESIS_756f45e3...	0	1435 days 18 hrs ago	GENESIS	→	0x756f45e3fa69347...	200 Ethe
GENESIS_f42f9052...	0	1435 days 18 hrs ago	GENESIS	→	0xf42f905231c770f0...	197 Ethe
GENESIS_2489ac1...	0	1435 days 18 hrs ago	GENESIS	→	0x2489ac126934d4...	1,000 Et
GENESIS_ddf5810a...	0	1435 days 18 hrs ago	GENESIS	→	0xddf5810a0eb2fb2...	17,900 E
GENESIS_c951900...	0	1435 days 18 hrs ago	GENESIS	→	0xc951900c341abb...	327.6 Et
GENESIS_6806408...	0	1435 days 18 hrs ago	GENESIS	→	0x680640838bd07a...	1,730 Et
GENESIS_9d0f347e...	0	1435 days 18 hrs ago	GENESIS	→	0x9d0f347e826b7dc...	4,000 Et
GENESIS_9328d55...	0	1435 days 18 hrs ago	GENESIS	→	0x9328d55ccb3fce5...	4,000 Et
GENESIS_7e7f18a0...	0	1435 days 18 hrs ago	GENESIS	→	0x7e7f18a02eccaa5...	66.85 Et

We also receive information on who was the miner, the reward for mining this block and the uncle rewards. Finally, there is the difficulty of the mining (as Ethereum started out with a proof of work consensus protocol), size of the block, the gas used in the block and the limit, extra data that can be added by the miner, the hash from the block header from the previous block, the parent hash, the SHA3uncles, and the Nonce. Clearly, there is a lot more information included in the Ethereum blocks when we compare these with the Bitcoin blocks. Figure 3.1 gives us an initial idea of how the block headers look like (based on the Ethereum Yellow paper).[227]

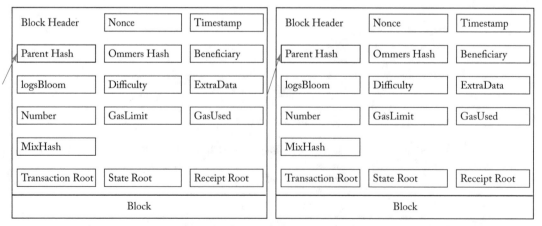

Figure 3.1: Block header.

227 Wood, G. (2019) Ehtereum: A secure decentralized generalized transaction ledger byzantium version. https://ethereum.github.io/yellowpaper/paper.pdf.

Several of these parameters we have discussed before when we were looking at the genesis block of the Ethereum blockchain. This information is based on the information given by Gavin Wood in his Yellow paper. There are three Merkle Patricia Tree roots in the block headers: transactions root, state root, and receipts root. The transactions root is, as you might have expected, the root of the Merkle Patricia Tree, linking together the transactions in the block. The state tree represents the entire state after the block has been processed. The receipts tree is a little bit more difficult to understand. It represents the "receipts" from the transactions. It is itself made up out of four separate items: medstate, gas_used, logbloom, and logs. In short:[228]

1. the medstate is the state root after the transaction has been processed;

2. the gas_used is the gas used after processing the transaction;

3. the logs is a list which again consists out of some other elements: logger's account address, a certain amount of topics and data that are produced by the LOG0 .. LOG4 opcodes during execution of the transaction; and

4. the logbloom is a bloom filter made up of the addresses and topics of all logs in the transaction.

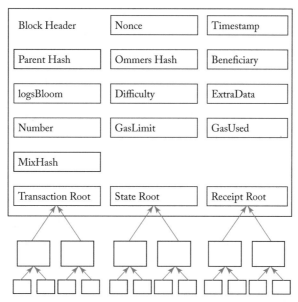

Figure 3-2: Ethereum block header.

Other information that is stored in the block header is:

- the parent hash is the hash of the parent block's header (using keccak256);

228 https://github.com/ethereum/wiki/wiki/Design-Rationale.

- the nonce is a 64-bit hash that, when combined with the mixHash proves that there was enough computation used to actually mine the block;

- the beneficiary is the successful miner of the block that will receive the fees from mining the block;

- the timestamp which is the Unix timestamp of the block;

- the logsBloom which is the Bloom filter on the logs;

- the difficulty which shows at which difficulty level the block was mined;

- the extraData which has in it… you guessed it: extra data;

- the number is the block number of the block, starting from 0 for genesis;

- the gasLimit which represents the gas (explained later) limit per block; and

- the gasUsed which is the sum of total gas used by the transactions in the block.

And then we still have one value to explain: the ommershash. This is the hash of the ommers in the block. So what the hell are ommers? In short, these are the stale blocks that arise in the network because of the high block time. They are called ommers or uncles and have the same grandparent as the current block that is being mined. Because of the GHOST protocol, the miners of these blocks also receive a reward, which is smaller than the reward for the miner that was able to mine the full block but still provides enough incentive to follow the main chain. What else can be found in the blocks of the Ethereum blockchain except for the block header? Well there is the set of transactions of course but also the other block headers of the current block's uncles.

3.4   GHOST PROTOCOL

The Ethereum network also makes use of the GHOST protocol because of the high block time of the network. This fast processing time leads to a lot of stale blocks or "uncles" which are now used within the calculations of the network to prevent too many forks. It is a limited version of the GHOST protocol that is being used by Ethereum though, to prevent high calculation costs. Uncles can only be included up until the "7th generation." On top of that, unlimited GHOST would allow miners to mine any block and still claim reward, pushing them away from the main chain. So, what is the design rationale behind all this? The goal of the network is to have a block time of 12 sec. We have learned before however, that this is in reality a bit longer due to network latency. On top of the 7th generation limitation, there is also a 1 block descendant limit. This both for security and simplicity reasons.

Interesting to know is that when it comes to the validation of the uncles, only the headers are checked. This is done to make the process lighter and easier but incorporates of course limited risks.

3.5   UTXOS?

In the section on Bitcoin I talked quite extensively about UTXOs and why they are of such a major importance in the network (and in many other networks that use the same mechanism). Ethereum chose to leave this approach and wanted to make use of accounts. These accounts consist out of a balance, and in some cases out of a code and internal storage. If you wish to make a transaction, it is simple: your account is checked and if the account has enough balance on it, the account is debited for the payment you wish to make while the receiver has his account credited. If there is a code, this code can be executed based on the result of the transaction, and even the internal storage can be adapted as a result of the transaction. Discussions on what is the best strategy for dealing with transactions are manifold but the main points are these: (1) UTXOs provide better privacy when it comes down to the transactions you are making and (2) it can give you a certain advantage when we are talking about certain scalability paradigms (but both are arguable when you wish to make use of smart contracts and dApps, and even so, in the past it has been proven that frequent users of the Bitcoin network could be identified, whether they liked it or not). The advantages of accounts are given by simplicity as it is easier to understand, greater fungibility as the source of the coins is no longer a blockchain-level concept, and it is easier for light clients to access all data related to an account when they simply need to look for the account in the state tree. Finally, the main advantage is the space saving compared to UTXOs both in storing information (account has one truth vs UTXOs which need to be combined to make a transaction work) and the transactions can be smaller as you again, only need one input, one output, and one signature. In the world of Ethereum, there are two types of accounts: externally owned and contract accounts. The first are owned by owners with a private key, just like you and me (in case you have an Ether account). These can be freely used to send transactions to other externally owned accounts or to contract accounts when the correct private key is used to sign the transactions. The contract account has code associated with it and is controlled by contract code. This means that it can only be activated when it actually receives a transaction. So, what comprises the "account state"? There are four components that are essential: the nonce, the balance, the storageRoot, and the codehash. The nonce is the number of transactions actually sent from the eternally owned account or the number of contracts created by the account in case it is a contract account. The balance is the number of Wei owned by the address (details later). Next, there is the storageRoot which is the Merkle Patricia tree root of the storage contents. Finally, there is the codeHash which is the hash of the EVM code for the account.

3.6   TRANSACTIONS

Just as with the Bitcoin blockchain, also here we are dealing with transactions and messages that are being send throughout the network. We can basically define two types, message calls and contract creations, but both have the same basic structure.

- The nonce is the number of transactions send by the sender.

- The gasPrice is the price the sender is willing to pay for the execution of the transaction.

- The gasLimit is the maximum amount of gas the sender is willing to pay.

- The "to" is the recipient-address.

- The value is the amount of Ether/wei that is to be transferred.

- The "v", "r", and "s" are used to generate the signature of the sender.

- The init-field is only included when we have a contract-creating transaction which gives us an EVM fragment that is used to initiate the new contract account.

- The data field is only used for message calls and contains the input data for the transaction.

We know from before that contract accounts can only be initiated when they receive a transaction from an externally owned account. This is not entirely true as in some cases contract code, once initiated, can call other contracts by making use of internal transactions.[229] Important for developers to know is that these subsequent transaction don't have a gas limit which means that the initial transaction should contain enough gas to power the first transaction and all subsequent transactions that follow. There are some other rules to consider when we talk about transaction execution in the Ethereum network environment. Within the transaction there must be a valid signature and a valid nonce (the counter if you remember). There are also validity checks on the intrinsic and upfront gas costs (defined later in this chapter) before we can move on to the actual execution. During execution there will always be a so-called "substate" that tracks the changes in the state while the transaction is executing. This substate tracks the following items:

- the "refund balance," which contains the amount to be refunded to the sender;

- the "log series," which are a number of logs referring to the EVM execution; and

---

[229]   Kasireddy, P. (September 27, 2017). How does Ethereum work, anyway? *Medium.* https://medium.com/@pree-thikasireddy/how-does-ethereum-work-anyway-22d1df506369. Accessed December 13, 2019.

- the "self-destruct set" which is a set of accounts that will be deleted once the operations are finished. Refers to the self-destruct opcode that can be used to delete a contract and send the remaining gas to the senders account.

3.7   SERIALIZATION

We learned before that the transaction has to follow a specific format called "Recursive Length Prefix" (RLP), which is the serialization format used in Ethereum. The RLP function takes in an item or a set of items and follows a specific encoding procedure.[230]

- If the input is a single byte in the range 0x00 – 0x7F, we know that byte is its own RLP encoding.

- If the input is a special byte, the encoding starts with 0x81.

- If the input is a string between 2–55 bytes long, the encoding scheme consists of 0x80 + the length of the string in bytes followed by the hex value of the string.

- If the input is a string longer than 55 bytes, the encoding scheme consists out of 3 different parts. We end with the hex value of the string upmost to the right, in the middle we have the length of the string in hex, and to the left we have 0xb7 + the length of the middle value.

- If the input is a non-value, the RLP encoding is 0x80.

- If the input is an empty array, the encoding is 0xc0.

- If the input is in the range 0x80 – 0xFF, the RLP encoding concatenates 0x81 to the input.

- If the input is a list with a total payload between 0 and 55 bytes, the RLP encoding starts with 0xc0 + length of the payload + the hex value of the items.

- If the input is a list with a payload longer than 55 bytes, the encoding scheme consists out of 3 different parts. We end with the hex value of the items upmost to the right, in the middle we have the length of the payload in hex, and to the left we have 0xb7 + the length of the middle value.

230 Chinchilla, C. (August 2, 2019). RLP. *Ethereum Wiki*. https://github.com/ethereum/wiki/wiki/RLP. Accessed December 13, 2019.

The decoding of the RLP-encoded transactions will always start with the first byte as this will make immediately clear which data types are being handled. The decoding reverses the rules which were set before and make clear what needs to happen and what information is being transferred.

3.8   SIGNING

To sign a transaction, one can call the "eth_sign" method from an Ethereum client which makes use of the keccak256 hashing algorithm. It looks like this: eth_sign(keccak256("\x19Ethereum Signed Message: \n" + len(message) + message)). As mentioned before, there is a nonce included in the transaction once it is being sent to prevent replay attacks from attackers that are able to capture our messages. The elliptical curve digital signature algorithm being used makes use of the parameters in the transaction called r, s, and v to create the actual signature. r and s are both 32 bytes and concatenated together to form the first part of the signature. The v is the 65th byte of the signature itself. To be able to verify the signature, these have to be split out and this way the sender's address can be recovered. In case the transaction was tampered with, the sender's address will no longer be correctly identified. On top of that, also the message needs to be verified, it can perfectly be that the signature is correct but that the message has been altered. The message is recreated by making use of the same parameters that should have been used and verify if the outcome is similar to what has actually been sent.

3.9   ETHER, FEES, GAS, AND FUEL

The cryptocurrency linked with Ethereum is the coin called "Ether." Opposite to the Bitcoin network, the main goal of the network wasn't the creation of the cryptocurrency but rather Ether is an enabler for the development for smart contracts and DApps on the network. It is used to power the applications and prevent certain forms of abuse while at the same time making sure that the miners and participants get paid. Another major difference is that there is no hard cap on the creation of Ether which makes it a coin that is much more susceptible to inflation. We have seen clear evolutions over time where there have been inflation rates around 14% in the years 2015 and 2016, which decreased to a steady 7.4% in 2018.[231] To give you an idea, Bitcoin has a yearly inflation rate of about 4.25%. However, when the Casper protocol will be implemented, it is suspected that the inflation rate will decrease to as low as 0.1%. If you have ever worked with the Ethereum platform and network before or have at least read about it, you will have certainly seen the word "gas" or "fuel." But what is it? As we already shortly described above, gas was invented to deal with several problems that Turing complete machines might run into. One of these is the halting problem which can lead

[231]   Conner, E. (July 27, 2018). A case for Ethereum block reward reduction to 2 ETH in Constantinople (EIP-1234). *Medium.* https://medium.com/@eric.conner/a-case-for-ethereum-block-reward-reduction-in-constantinople-eip-1234-25732431fc77. Accessed December 11, 2019.

to the infinite loop but also DoS-attacks and others are mitigated by the use of this new concept. If we are talking about "simple" transactions, gas can be seen as the transfer fee that must be paid for your transaction to take place. It is also the reward that miners get for mining your transaction into a block. But as you know now, a lot more can take place on the Ethereum platform, so what does gas measure when it comes to executing smart contracts? Gas is the computational effort that it will take to execute all the operations that are enclosed within a smart contract.[232] Making use of the EVM takes up a lot of effort and should therefore only be used for simple tasks and the fees help encourage all participants to keep their computations to a minimum. The denomination for gas can be found in the Table 3.2.

Table 3.2: Gaws denomination		
Unit	Wei Value	Weis
Wei	1 Wei	1
Babbage	1e3 Wei	1,000
Lovelace	1e6 Wei	1,000,000
Shannon	1e9 Wei	1,000,000,000
Szabo	1e12 Wei	1,000,000,000,000
Finney	1e15 Wei	1,000,000,000,000,000
Ether	1e18 Wei	1,000,000,000,000,000,000

The actual transaction fee is calculated based on a simple formula: gas limit × gas price = max transaction fee. The limit is set by the sender and represents the maximum amount he is willing to pay for the transaction to be executed. If there is enough Wei to be found on the account of the sender, the transaction will be executed otherwise the transaction is invalid. This fee will eventually end up as a reward for the miner that mines the block which includes the transaction that was executed by the sender. There is also a check on this gas limit when the transaction is created. The gas limit should be equal or greater than the intrinsic gas used by the transaction. This intrinsic gas consists out of 21,000 gas + 4 gas for every byte of data equal to 0, or 68 gas for every byte of data ≠ 0. On top of this an additional cost of 32,000 gas is possible when we are dealing with a contract creation transaction. A final check is performed on the account of the sender by adding the maximum transaction fee and the value of the transaction together. This is called the upfront gas price and the account of the sender should contain a high enough balance to cover these costs. If this is the case, the transaction can move to the next steps of execution. Similarly, you can store data on the Ethereum blockchain but also for these services you have to pay. Here the fee is calculated based on the smallest multiple of 32 bytes used. If you execute an operation that has as aim to free up space, you don't have to pay a fee for this transaction and even a refund is given for the freed up space.

[232] Rosic, A. (2018). What is Ethereum gas? *Bockgeeks*. https://blockgeeks.com/guides/ethereum-gas/. Accessed December 11, 2019.

3.10   THE MILESTONES OF ETHEREUM

The development of the Ethereum platform has followed several prototypes which had each their own code names. Below you can find an overview of the several versions of the platform that have already been created so far. The first really stable version of the network came with the Homestead update which brought several improvements to the network.[233] Serenity is supposed to be the final version of the network with the implementation of Casper 2.0.

Table 3.3: Ethereum platform versions		
Version	Code name	Release date
0	Olympic	May, 2015
1	Frontier	30 July 2015
2	Homestead	14 March 2016
3	Metropolis (vByzantium)[234]	16 October 2017
3.5	Metropolis (vConstantinople)[235]	28 February 2019
4	Serenity	TBA

Just as with the Bitcoin network, also Ethereum has seen its protocol updates over time and its subsequent hard forks (later more). The Ethereum network differs in several ways from the Bitcoin network. One of these differences, is the use of accounts and balances (called state transitions). So unlike Bitcoin, it does not have to rely on UTXOs. Another major difference is that the block time is only 15 sec (compared to 10 min for Bitcoin). The return of ether for mining a block is consistent, and this is why one sometimes refers to Ether as the oil of the cryptocurrencies. There is also a major change ongoing in the way the protocol works as it is moving from the proof of work protocol to the proof of stake protocol. It does this in several implementations, to make sure that the change in the network is not too significant. The Ethereum also makes use of a concept called "gas," which is measured in Gwei. This is the transaction fee one has to pay based on the computational complexity, bandwidth being used and the storage needs.

3.11   THE STAGES OF ETHEREUM EXPLAINED

As with the development of Bitcoin and other blockchain platforms, there have been several stages in the development of Ethereum. The difference with the Bitcoin platform was that the development of Ethereum has been clearly staged from the very beginning. Some of these steps were not planned beforehand but the main timeline was always clear. I would like to point out that not everything described below will be immediately clear but will be explained on the next pages. So,

---

[233]   Improvements to transaction processing, gas pricing and security among others.
[234]   This was a soft fork which reduced the complexity of the EVM and added support for zk-SNARKS.
[235]   This update was a hard fork.

no panic, everything will become clearer when you move along and don't hesitate to refer back to this section on the timeline. It will make clear what is already there and what will be added in the near future.

3.12   FRONTIER

This was the first stage of the Ethereum platform and started on July 30, 2015 and lasted until March 2016. To be completely honest, before the release of Frontier there was another version that was released called "Olympic." It was the final pre-release version that was opened up to the world so that the limitations of the network could be tested by willing participants.[236] There was the focus on transaction activity, virtual machine usage, mining prowess and general punishment. It lasted for about 14 days before the launch of the Frontier 1.0 version of Ethereum. The Olympic testnet was replaced with the release of Frontier with "Morden" which was a Frontier-equivalent testnet. The mining on this first version of Ethereum could start once the hardware was installed and the genesis block could be generated. As it was the first version of the network software, the network was plagued with initial bugs and updates propagated through the network to adjust for these issues. There was also the implementation of so-called "canary contracts," which were only used in this version of the platform. It was a centralized check that was created to follow up on the frontier clients and possible consensus issues. There were four "switches" that could be either one or zero and these were controlled by the Ethereum developers. From the second that two of these were turned on, the mining would stop and the participant was forced to update their client so that these wouldn't prevent a chain upgrade. The used consensus algorithm was the proof of work algorithm called ETHhash even though the developers already knew they wanted to move to a proof of stake protocol in the future.

3.13   ICE AGE

This stage started at block 200,000 (September 7, 2015) and introduced a time bomb that would increase the difficulty in the proof of work protocol used by the network. The name "ice age" refers to the "freezing" of the blockchain because of the increased difficulty, the fact the miners wouldn't be able to keep up and as a direct effect that the block time would increase. The main reason was to stimulate the participants in the network to move to the proof of stake protocol when it was ready to be implemented. This means that at the moment the network is ready to implement the new proof of stake "Casper" protocol (the Serenity update), there will be a hard fork with the remaining proof of work network which would force every miner to switch to the new network. Why? Remaining on the old network would leave the participants on a PoW network where the difficulty

---

[236]   Buterin, V. (May 9, 2015) Olympic: Frontier pre-release. *Ethereum Blog*. https://blog.ethereum.org/2015/05/09/olympic-frontier-pre-release/. Accessed December 17, 2019.

would eventually become so high, that the participants could no longer mine. Stephen Tau stated that the bomb would go off at block 200,000 but that the results would only be noticeable after about a year.[237] After this time, the difficulty would increase significantly, leading to higher block times. The goal was that the Serenity update would be ready by this time. It seems clear now that this timeline wasn't met and the difficulty bomb has been delayed several times already to prepare for the Serenity update. The algorithm behind the time bomb looks like this:

Block_diff = parent_diff + parent_diff // 2048 * max(1-(block_timestamp − parent_timestamp) // 10, -99) + int(2**((block.number // 100.000) − 2) with // the division operator resulting in 8 // 4 = 4 and 9 // 4 = 4. [238]

3.14   HOMESTEAD

The next stage started at block 1,150,000 (March 14, 2016 or Pi-day) and was the second implementation of the Ethereum network. It was a hard fork because it brought protocol changes that were not backward compatible. Similar to the Bitcoin network, also in the Ethereum world there are "Ethereum Improvement Proposals" (EIPs). Several of these were implemented with Homestead such as EIP 2 which increased the cost of contract creation from 21.000 to 53.000.[239] In case there is no gas enough to create the contract, it simply fails (instead of leaving an empty contract). Also, transaction signatures are now checked. There are also EIP 7 and EIP 8. The first introduced a new opcode called DELEGATECALL which can be used for contracts that create contracts but don't repeat additional information. The second changed the RLPx discovery protocol and RLPx TCP transfer protocol so that clients could deal with future network upgrades (before the client would simply stop communication).

3.15   DAO

Later in this chapter explained in more detail, around block 1,192,000 a hard fork took place when there was a hack in the DAO. Discussions followed, after which a part of the community decided to refund the victims of the hack. A part of the community would remain in the old chain, creating "Ethereum classic."

---

[237] Tual, S. (August 4, 2015) Ethereum protocol update 1. *Ethereum Blog*. https://blog.ethereum.org/2015/08/04/ethereum-protocol-update-1/. Accessed December 17, 2019.

[238] Rosic, A. (2017). What is Ethereum Metropolis: The ultimate guide. *Blockgeeks*. https://blockgeeks.com/guides/ethereum-metropolis/. Accessed December 18, 2019.

[239] https://ethereum-homestead.readthedocs.io/en/latest/introduction/the-homestead-release.html.

3.16   THE DAO ATTACK

It was in 2016 that a major event occurred which would influence not only the perception of the Ethereum platform but also the concept of the "decentralized autonomous organization'\." There was a fundraising project that made use of a decentralized autonomous organization, called "the DAO" (real original, I know). The DAO was created by Christoph Jentsch together with his brother Simon. Their goal was to fund projects using ether via the DAO. It was a very popular organization and at the time (May 2016) it had attracted almost 14% of all ether that had been issued to that date. The idea was that the members could vote on which projects could win the investments and several security features were built in the decentralized autonomous organization to prevent abuse of these voting rights.

However, darkness loomed and soon an important security vulnerability was discovered in the code supporting the organization. This flaw was pointed out by several people and in June several solutions had been proposed to fix the recursive call issue but it would be too late. It was June 17, 2016 when an attacker exploited several vulnerabilities and transferred 3.6 million Ether from the DAO to an account that had a 28-day holding period. Following this event, the community got torn over the question whether a hard fork should occur so that the lost funds could be returned. With no consensus in the community, a hard fork occurred with Ethereum continuing on this fork, while the original chain remains to be used by what is now known as Ethereum Classic. Ethereum Classic also had to deal with the time bomb that was part of their code and hence the Ethereum Classic chain had to perform a hard fork themselves to get rid of this feature. Eventually the tokens related to the DAO would be de-listed by the end of 2016.

3.17   TANGERINE WHISTLE

This hard fork was implemented to address the gas calculation in I/O-heavy operations and to address the results of DoS attacks so that it addresses the immediate network health issues which were the result from previous attacks. The idea was that the opcodes that read the state tree were in fact underpriced. These opcodes were easy to add to smart contracts but difficult to process for the clients, which lead to delays in the network. We were at that moment at block 2,463,000 and the change was based on EIP 150.[240]

3.18   SPURIOUS DRAGON

Starting at block 2,675,000 other defense mechanisms were implemented to address other DoS-attacks and replay attacks and hence the next hard fork took place on the November 22, 2016. EIP 155 (replay attack protection), 160 (EXP opcode cost increase), 162 (state trie clearing to further

---

[240] http://eips.ethereum.org/EIPS/eip-608.

help prevent DoS-attacks), and 170 (the contract code size limit was brought to 24,576 bytes) were all implemented with this update. [241]

3.19 METROPOLIS BYZANTIUM

The next stage in Ethereum development was called Ethereum Metropolis Byzantium and occurred at block 4,370,000. It brought to Ethereum the introduction of zk-SNARKs to improve privacy on the blockchain. Next, the time bomb also was delayed another 18 months to allow the developers more time to work on the next implementations and versions of the Ethereum network. Finally, the implementation of smart contracts has been made even more flexible and robust. Specifically, if a contract cannot move on to the next "state" during execution because of a shortage of gas, the contract is reversed to the previous state without spending all the gas. On top of that, the RE-TURNDATA opcode is added so that variable length values can be returned. Finally, the upgrade of the network wants to achieve account abstraction, making the network more user friendly so that users will no longer need technical knowledge to make use of the platform.

3.20 METROPOLIS CONSTANTINOPLE

The next phase in the Ethereum timeline is Metropolis Constantinople. It was normally planned for the beginning of 2019 but has been pushed forward. Several reasons can be thought of for this push toward the future but one of them was the fact that an auditing team discovered that the upgrade contained a vulnerability that was linked with EIP 1283 that would introduce cheaper cost of storage which in turn introduced the reentrancy attack. What does this mean? Easily explained, an external contract might communicate with a smart contract. If the external contract is in any form malicious, it might try to take control of the smart contract's code and make unexpected changes. Particularly, it can try to re-entering in a particular spot in the code, withdrawing Ether from the smart contract (withdrawBalance() function is abused here). It would have affected only a small number of smart contracts, but still, the developers are now working on the problem to make sure that no vulnerabilities are introduced. It was also this type of attack that was used in 2016 on the DAO (see earlier). The following EIPs would be implemented during this hard fork: EIP 145, 1052, 1283, 1014, and 1234. The first introduces bitwise shifting which allows code (and the underlying operations) to become better optimized so that the code can be processed faster and at lower cost. Next, EIP 1052 introduces a new opcode EXTCODEHASH which returns the keccak256 hash of contract code. This can be interesting for contracts which have to check the bytecode of other contracts without actually using it. Again, this will lead to efficiency and lower cost. EIP 1283 refers to the SSTORE opcode which can be used for gas metering. Again, the aim is to reduce costs for

---

[241] Jameson, H. (November 18, 2016). Hard Fork No. 4: Spurious Dragon. *Ethereum Blog.* https://blog.ethereum. org/2016/11/18/hard-fork-no-4-spurious-dragon/. Accessed December 18, 2019.

developers.[242] The next EIP introduces state channels (explained in more detail below) which are similar to the Lightning network that we find in the Bitcoin world. Finally, there was the plan for the transition to the proof of stake consensus protocol with the introduction of Casper FFG (explained in detail below). However, this plan was cancelled late 2018 where the developers eventually decided to move away from Casper FFG and immediately go for the full implementation.

3.21   SERENITY

Serenity is the final stage of the Ethereum network development and is also known as Ethereum 2.0. One of the main goals of this final upgrade is to create a more scalable and efficient platform that is capable of handling thousands of transactions per second. Several implementations are expected divided over phases, such as the beacon chain without shards (phase 0), shard chains without EVM (phase 1), the implementation of a new execution engine (phase 2), phase 3 with the light client state protocol, phase 4 will bring cross-shard transactions, phase 5 tight coupling with main chain security, and finally phase 6 with upper-quadratic sharding.[243] This all will eventually lead to "Ethereum 3.0," which will contain several new implementations.

3.21.1 PHASE 0

In a little more detail, phase 0 brings the implementation of the beacon chain without sharding and validators create an RNG via RANDAO in block proposals, organize into proposers and attestation committees based on the RNG output, and create crosslinks for stubbed shards. The beacon chain will be the proof of stake based blockchain version of Ethereum. So far, the Ethereum network has always made use of proof of work but with the beacon chain this is supposed to change with the introduction of Casper protocol. The beacon chain will also be used for sharding in the later phases of Serenity. Validators will have to put up a stake of 32 ETH to be able to join the process. These validators will be organized into committees so that they can vote on the proposed blocks. These committees and their validators will make use of "attestations" to vote on these proposed blocks (beacon blocks and shard blocks). All of this will be done by the beacon chain. On the beacon chain there will also be a new ether called "ETH2," which is a new asset to be used by the validators. In this phase there will be no way yet for the participants to withdraw this new currency. Finally, RANDAO will introduce sufficient randomness into the system when it comes to organizing validators in proposers and committees. [244]

[242] Mitra, R. (2019). Understanding Ethereum Constantinople : A hard fork. *Blockgeeks*. https://blockgeeks.com/guides/ethereum-constantinople-hard-fork/. Accessed December 20, 2019.

[243] Ray, J. (March 4, 2019). Sharding roadmap. *Ethereum Wiki*. https://github.com/ethereum/wiki/wiki/Sharding-roadmap. Accessed December 20, 2019.

[244] https://docs.ethhub.io/ethereum-roadmap/ethereum-2.0/eth-2.0-phases/.

In this phase the proof of work blockchain that has always been used by the Ethereum blockchain environment will coexist with the beacon chain and all transactions and smart contract computations will still take place on this "old" chain.

3.21.2   PHASE 1

Phase 1 starts with shards (without chain state execution or account balances) and binary large objects (or BLOBs) are collated in shards without transactions. Also, notaries will see the light of day in this phase. Sharding is the solution that has been put forward by Ethereum to deal with the scaling problems that are currently plaguing public blockchain implementations. The problem relies on the fact that a public blockchain cannot be at the same time decentralized, secure, and scalable: a choice needs to be made between two of the three. The Bitcoin network tries to solve this with implementations like the Lightning network or sidechains that sparsely interact with the main chain. Sharding leads us to a completely new approach. Before the implementation of sharding, each node has to process each and every single transaction from the network. This again leads to the network only being as fast as the individual nodes.[245] Therefore sharding allows the entire state of the network to be divided in so-called "shards," which represent each their own piece of state. Each of these shards would be linked to the beacon chain by making use of the Merkle trees (combined data roots) creating a connection between the two which are also called "crosslinks." Once such a block with a "combined data root" has been accepted on the beacon chain, the other shards know they can rely on it for cross-shard transactions. Each of the shards will store receipts of each transaction so that they can still communicate with the other shards and perform transactions with each other. One of the issues is that sharding is way easier to implement when the network makes use of proof of stake instead of proof of work. Active validators can just be assigned to different shards.[246] This is why the Ethereum developers are working on sharding as a part of the Casper protocol implementation (explained below). In short, basically it means that the main chain is chopped into smaller chains where the node is acting as a full node for a certain shard and as a light client for the other shards. Similar to phase 0, the proof of work chain and the beacon chain with the shards will friendly coexist with one another in this phase as well. This means in practice that validators and miners both will receive rewards for the time being, leading to the necessary inflation but it is during this time that the proof of work chain should lose its appeal while participants are crossing to the beacon chain.

[245] Jordan, R. (January 10, 2018). How to scale Ethereum: Sharding explained. *Medium.* https://medium.com/prysmatic-labs/how-to-scale-ethereum-sharding-explained-ba2e283b7fce. Accessed December 21, 2019.
[246] https://education.district0x.io/general-topics/understanding-ethereum/ethereum-sharding-explained/.

3.21.3   PHASE 2

Phase 2 will introduce structured chain states to the shards combined with the use of smart contracts. This will bring with it also accounts, contracts, states, and all other basic concepts that we are already using on the Ethereum network today. Perhaps most important to understand is the introduction of eWASM or Ethereum-flavored WebAssembly. WebAssembly (or Wasm as a contraction) is a new, portable, size-, and load-time-efficient format. WebAssembly aims to execute at native speed by taking advantage of common hardware capabilities available on a wide range of platforms and WebAssembly is currently being designed as an open standard by a W3C Community Group.[247] All of this means that it is a virtual machine within the computer that can optimize the execution of commands and operations. It does this by converting or immediately executing of commands because it has a knowledge of the hardware it is running on. The introduction of eWASM is another answer to the question of scalability and is aimed at EVM. Currently, when EVM has to compile code, every node has to compile the node. This is not only very costly but also limits the speed of Ethereum to the speed of the Ethereum Virtual Machine. In this specific implementation WebAssembly is being designed exclusively to work with the smart contracts that exist in Ethereum. It is the goal that eWASM will be able to replace the EVM, optimizing how code is run in the Ethereum network and dramatically improving the transaction throughput in the network. On top of that, it can be more secure as it is standardized and it delivers support for more programming languages, leading to a broader developer base. Finally, there would also be the implementation of state rent so that developers would have to pay for eWASM storage over time to make sure that unused data is removed in a timely fashion.

3.21.4   PHASE 3

Phase 3 (and the other phases) are still very speculative. I have added the information here but as the previous steps and phases already have shifted and changed a lot, these steps are even more speculative). This phase follows with the introduction of state minimized executions. This was the focus of light clients because the previous phases were all focused on full nodes. How this would be implemented is still under discussion.

3.21.5   PHASES 4, 5, AND 6

Phase 4 comes with cross-shard transactions. The main introduction that comes with phase 5 is the introduction of fork-free sharding. Finally, there will be phase 6 which, as we mentioned before, comes with super-quadratic sharding. All of these implementations are aimed at increasing the scalability of the network while adhering to the security that we would like to see in the network.

---

[247]   https://github.com/ewasm/design.

3.21.6   ETHEREUM 3.0

So what is Ethereum 3.0? The first lines put forward talk about the integration of zk-STARKs in the network and heterogeneous sharding.

3.22   SCALABILITY AND THE CASPER PROTOCOL

So far we have been focusing on how the network works and how it will be changed in the future but we still have some concepts to explain. State channels and plasma are two projects that also focus on the scalability of the network and aim to improve it for the participants. Finally, I also added an explanation on the Casper protocol.

3.22.1   STATE CHANNELS

State channels are one of the several solutions put forward to deal with the scalability issues that Ethereum is facing. State channels are very similar to the Lightening network that is being used by the Bitcoin platform, allowing for off-chain transactions to take place, which can be propagated later when the channel is being closed. The state channels in the Ethereum world also allow for something else: they support state updates (hence the name). Comparable to the Lightening network, a certain amount of Ether is locked by sending the amount to a multisignature smart contract which can both accept and pay out the coins. Once the Ether is in the contract, the participants can sign transactions (of which each party contains a copy) that each have nonce to keep track of the chronological order. The channel can eventually be closed by emitting a transaction to the Ethereum main chain.[248]

3.22.2   PLASMA[249]

Plasma is yet another solution put forward to deal with the scalability issues of the Ethereum network. With this implementation it allows the creation of child chains that use the main chain (or a shard of the main chain) as a trust and arbitration layer.[250] Plasma allows for the creation of chains that can be used for specific implementations that currently aren't feasible with the existing main chain. These chains can be adapted when looking at block size, consensus algorithm, block times, and so on. Of course there are some limitations as there is still a need for a consensus algorithm that enforces the Nakamoto consensus incentive and a bitmap-UTXO commitment structure to enforce the state transitions. This way DApps can be created that fit any purpose depending on the needs of the participants, and increasing the scalability of the Ethereum network tremendously. How is all

---

[248] https://education.district0x.io/general-topics/understanding-ethereum/basics-state-channels/.
[249] http://plasma.io/.
[250] https://education.district0x.io/general-topics/understanding-ethereum/understanding-plasma/.

of this linked together? By making use of "plasma contracts" that connect to the root chain. These allow the transfer of assets between the main chain and the child chains. The general rule is that these assets have to be first created on the main chain before they can be moved to a child chain, this to prevent malicious activity from the child chain to propagate to the main chain. Another possible problem is the centralization on the child chains, which could lead to mined blocks that do not represent true transactions. Plasma has a solution for this: the possibility for each participant to show fraud proofs (by making use of a MapReduce computing framework). The main concern with plasma is that it takes quite a long time to withdraw assets (between 7 and 14 days).

3.23   THE CASPER PROTOCOL[251]

The Casper consensus protocol is a hybrid of the proof of work and proof of stake protocols. While the first is deemed undemocratic and is very costly in terms of hardware and energy. On the other hand, the proof of stake protocol is efficient and more secure but has the "nothing at stake" problem to deal with. Enter: the Casper consensus protocol. Casper differs from other proof of stake protocols as it punishes malicious actors in the network. We again work with validators that have to stake a portion of their Ether to enter the position. The validators are starting to validate blocks and when they have discovered a block that they deem to be a valid candidate to enter the chain, they have to place a bet on the block. If the block is accepted, the validators get rewarded for the bet they placed on the block, if the block is denied, they lose everything. Actually, there were two research projects undertaken by the Ethereum development team. On the one hand, there is the Casper the Friendly Finality Gadget (or FFG) and the Casper the friendly ghost: correct-by-construction (or CBC). The FFG version[252] was first proposed to aid the transition from proof of work to proof of stake. The proof of work protocol was still active, but instead every 50th block had a proof of stake checkpoint where the finality was assessed.[253] It provides extra finality over the standard proof of work protocol because there is total economic finality. About 2/3rd of the miners in the network put up their entire stake when validating a block, so they stand to lose a lot when they would try to act maliciously. There is also the possibility of a double finality, where in the case of Casper, the participants would have to choose a chain and the majority vote would select the main chain (resulting in a hard fork). This protocol was proposed several times but eventually was moved entirely from the implementation timeline.

The CBC[254] version will bring even more changes to the use of the protocol.[255]

---

[251] Rosic, A. (2017). What is Ethereum Casper Protocol? Crash course. *Blockgeeks*. https://blockgeeks.com/guides/ethereum-casper/. Accessed December 22, 2019.
[252] Also known as "Vitalik's Casper."
[253] Finality means that when a transaction (or any operation really), once done, is locked in the blockchain without the possibility to revert this. In reality, this can never be achieved for 100%.
[254] Also known as Vlad's Casper.
[255] Check out the presentation of Vlad Zamfir on https://www.youtube.com/channel/UCNOfzGXD_C9Y-MYmnefmPH0g/videos.

Table 3.4: CBC protocol design	
Normal Protocol Design	**CBC Protocol Design**
Formally specify the protocol	Formally but partially specify the protocol
Define protocol properties that must be satisfied	Define properties that the protocol must specify
Prove that it satisfies the properties	Derive the protocol so that it satisfies all the properties that it was stated to specify

If we want to derive the full protocol, we would have to implement an "ideal adversary" that would raise exceptions and list out any future failures that might happen.

3.23.1 SAFETY ORACLES

In this Casper V2 implementation, there has been a lot of confusion on how this might look like, also, due to the fact that the developer team behind the Ethereum network has several times changed the roadmap and the approach to this new protocol. In the latest update (at the time of writing) there was the following proposal. The Ethereum network would be split into several separate chains of which there would be 1 Ethereum proof of work chain, 1 beacon chain, and a number of sharding chains. The first is the current chain which still uses the proof of work protocol. If miners want to continue mining, they will have to deposit 32 Ether to the beacon chain, after which they will receive the roll of "validator." This beacon chain will become the main proof of stake chain within the Ethereum network and will also be the base layer of the sharding chains. It will link to these separate chains and make clear which blocks from these shards can be added to the main chain. This main chain will be the beacon chain. So what are these sharding chains? To prevent that every node will be working on every transaction, there will be a division in separate sharding chains, where nodes will be working on a specific subset of the transactions in the overall network. Simply put, validation and finality will be provided by the beacon chain while transactions and account data will be stored on the shard chains.

3.24   SMART CONTRACTS

We now had a very thorough discussion on how the Ethereum network works and what the blockchain looks like and often there was a specific term that kept on coming back: smart contracts. Smart contracts that can be executed on the network is one of the central concepts and most important addition brought by Ethereum. It allows for the execution of automated contracts when certain conditions are met. It is one of the main reasons why Ethereum is so popular and why there is an entire ecosystem of developers and independent projects that all refer to this main concept. An important concept to understand in the world of Ethereum and smart contracts is the so-called "halting problem" that keeps on coming back when you talk about Turing complete machines. It was a concept that was first stated by Alan Turing, and in short it means that it is impossible to create an algorithm that is capable of knowing if a program will actually terminate its execution for all possible inputs. In Ethereum this means that we cannot known if a smart contract will ever end. To solve this problem (but also other security related vulnerabilities), the concept of 'gas' was invented to deal with this directly. This way the execution of a smart contract will always be halted and why we say that smart contracts in Ethereum must be "terminable." This problem does also arise to other blockchain/distributed ledger technologies that aim to be Turing complete. The Bitcoin blockchain isn't Turing complete, so this issue doesn't arise as such in this environment. Every command is bound to finish in the Bitcoin environment. Furthermore, we state that smart contracts must also be deterministic, meaning that for a given input, the output must be the same every time. A final characteristic is that the smart contract must be "isolated." The smart contract and, more importantly, the results of a smart contract are isolated from the rest of the network. This is to prevent certain malware or issues from influencing the entire network and compromising the entire ecosystem. This is where the EVM comes in, as we explained before. Smart contracts are one of the novel innovations brought by the Ethereum network. The main purpose and meaning behind the term is that we are talking about a contract that can be executed without the intervention of a third party, but completely based on computerized transaction protocols. The idea here is that contractual obligations between several parties can be based upon computer code, which does not leave any room for interpretation but is executed from the second the necessary conditions are met for execution. This in combination with a blockchain network makes sure that the contract is replicated and stored while at the same time providing the necessary security and immutability. This concept was first brought to life with the advent of the Ethereum network but it is certainly no longer limited to Ethereum. A short distinction that we are going to make here is the difference between smart contracts and Ricardian contracts. If we follow the definition, a Ricardian contract is "a digital contract that defines the terms and conditions of an interaction, between two or more peers, that is cryptographically signed and verified. Importantly it is both human and machine readable and digitally signed." The main goal of a Ricardian contract is that it is a legally valid document that is stored in such a way that it can be executed by software. You can clearly see that there are certain points where both smart contracts and Ricardian contracts match but there are also points where they differ. A smart contract can be a Ricardian contract but certainly doesn't has to be and vice versa. If we can achieve both than that is nice, but in reality this will often not be the case. Now how are contracts actually created on the Ethereum platform? Well, in fact we need a completely new account for the contract so this needs to be created. This is done by following a specific set of steps:

- the nonce = 0;

- account balance = value that sender is sending with the contract creation transaction;

- storage = empty; and

- codeHash = hash of empty string.

The creation is finalized by the "init code" in the transaction which can in itself create more accounts, call other contracts, or send out some transactions.

3.25   BLOCKCHAIN ORACLES

Blockchain oracles are agents that reside on the blockchain to gather and verify real-world information and use this for the execution of smart contracts. This might seem a bit abstract but it comes down to this. A blockchain is a data structure and this is not capable of accessing data outside of its own network. Here the oracle enters the story as this is a third-party service that comes into play to provide the data necessary for a smart contract to be resolved. You might instantly spot a security issue. How do we trust the oracle? It is no part of the blockchain so there is no way of knowing if it is actually feeding us truthful data. Several techniques have now already been implemented to help build a security on top of these oracle. Notable examples are Oracalize with TLS notary-based proofs and Town Crier that makes use of Intel Software Guard Extensions. Furthermore, there are several types of oracles that we can identify. There is the difference between software and hardware oracles. The first provides us information from the online world, while the second feeds information from the physical world. We can also make a distinction between inbound and outbound oracles where the first feeds information to smart contracts, while the second sends information out of the blockchain environment. We will see later on several examples of blockchain oracles and how they interact with smart contracts and decentralized application on the Ethereum platform (and other platforms). They provide even more possibilities for the development of applications and new ways of working in a decentralized world.

3.26   DAPPS

The DApp or the decentralized application is the next step when we enter the world of blockchain and smart contracts. It adds another layer on top of smart contracts. This layer makes it a full-fledged application that can be used in a user-friendly way. For the user not that much changes. They can still work with a user interface with which they are familiar while at the same time they are dealing with a completely different underlying structure. Does this matter? In reality, this will only aid the adoption of blockchain based technology and applications, as users do not want to be confronted with the technological part of things, they just want to use applications for what they were intended. Nowadays, users do not want to go through lengthy processes of learning how something works, an (d)app has to be quick and easy to understand.

3.27   DECENTRALIZED AND AUTONOMOUS

With the introduction of smart contracts and decentralized applications comes the introduction of something different: the decentralized autonomous organization. This is an organization that is ruled by the code that has been imposed in the smart contracts that make up the organization. These rules are recorded and maintained on the blockchain together with all financial records. We already mentioned shortly in the timeline of Ethereum that there have been some issues in the past with the most famous one the hack of the DAO. Still, it is an interesting concept that is worth exploring as there are numerous future possibilities based on this a bit foreign concept. Several advantages can be thought of when this is implemented correctly. The need for third parties to approve and verify transactions taking place is removed and the code in place clearly defines the rules one has to live by when one wants to be part of the organization. You can solidify democratic voting systems and prevent fraud in a basic way, allowing all participants to aid in the future determination of the organization. The problem, of course, is the participation of all users in these votes. You could implement a system that forces participants to cast a vote but how far does the system remain democratic in that sense? Another challenge for the future is the legal status of such an organization. In the current legal frameworks, a decentralized organization would probably be considered a general partnership or a joint venture where all participants bare full legal responsibility. This means that all personal belongings of each party involved in a decentralized organization could be seized for the debtors of the organization.[256] Something to think about before you jump into unknown waters. Even more important is that the SEC in the U.S. has determined that this can be seen as unregistered securities offerings, which are illegal and can lead to prosecution.[257] It proves to be scalable, resilient, and decentralized governance, enabling broader adoption of DAOs by the public. DAOs have to deal with several problems, which are immediately also the reason why we haven't seen many of them out there as of yet. Daostack tried to solve these with their own interpretation of the problems and the possible solution. Scalability is an obvious problem as the participation of all members of a DAO for each decision is inherently unlikely, leading to a situation where the entire DAO is in an endless state of indecisiveness. While the opposite leads to misrepresentation of the real consensus within the DAO, or even worse: collusion and malicious decisions. A possible solution that is now in use is called holographic consensus, which is based on decisions made on a local level with limited attention and voting power.[258] It makes use of relative majority which means that the only majority needed is the majority provided by those participants that actually voted in the given timeframe, very much like modern democracies nowadays tend to work during elections.

[256]   Hinkes, A. (May 29, 2016). The law of the DAO. *Coindesk*. https://www.coindesk.com/the-law-of-the-dao. Accessed December 28, 2019.

[257]   https://www.sec.gov/news/press-release/2014-111.

[258]   Field, M. (November 12, 2018) Holographic consensus – part 1. *Medium*. https://medium.com/daostack/holographic-consensus-part-1-116a73ba1e1c. Accessed January 12, 2020.

Of course there is also a minimum amount of participants that actually need to participate before the vote can be seen as valid. As a lot of proposals are propagated throughout the DAO, proposers can actually give their proposals a value, so that it goes up in the collective attention. To motivate voters to actually vote, they need to become motivated and their efforts compensated. They get rewarded with the native DAO token. There is, however, also the need of another token. Proposals need to be filtered to protect the decision process and create a better executed open, economic, and permissionless network for the predictors. When proposals are accepted, the predictors are rewarded, if not, they lose part of their stake in the network.

3.28 WEB 3.0

With the coming of blockchain, the digital world has entered a new phase which is often called "Web 3.0," a term once coined by Gavin Wood and aimed at a completely new way of working and application building. This term has been popularized with the rise of decentralized applications and the support of the Ethereum community but it refers to a broader change and evolution of the internet. Critics state that it is nothing more than a marketing term to help people push to this new type of applications. In short, Web 1.0 was the "read-only web" meaning that you could look up information and read it but the fun stopped there. There are still a lot of websites that still follow this basic concept, although a new type of websites joined the world around 2002–2004 where users could also start adding their own content and upload information to websites. Social media applications are a prime example of how users are now influencing the world around them. Web 3.0 is only just the next natural step with the use of 3D, AI algorithms that will filter out the best data for the user combined with the Semantic Web that will be able to interpret the data and match even better the records that are registered on a website, therefore giving more information and even meaning to what can be found on the World Wide web. But it goes even further. In Web 2.0 you basically need to contract your financial institution to make a payment while Web 3.0 aims to create a world of payable machines and internet. For the user this will mean a completely new understanding of the internet as it is. Even though web pages might just appear as they do now, the possibilities that are being offered will increase vastly over time and the "power houses" that we know today will be confronted by an entirely decentralized opposition of web applications and developers that have adapted to a new way of working and creation. Ethereum has offered a new web stack for developers on how they can adapt to this new world and how they see the future of development but again, this is a much broader evolution that you should aim to understand. Below you can find the web stack as envisioned by Ethereum and others (largely based on an abstraction made by Stephan Tual in 2017).

Several of these components we are going to discuss below so that you can have a better understanding of how decentralized applications are created and what we can find in the several layers of the architecture. Some of these components have already been discussed in previous chapters so I will not repeat these here.

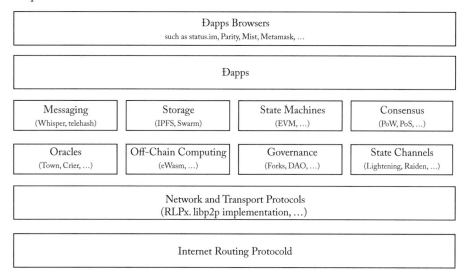

Figure 3.3: **Web 3.0 stack**

3.28.1 ETHEREUM WHISPER

Ethereum Whisper is a communication protocol which allows DApps to communicate with each other over the blockchain platform. It was called an "identity-based pseudonymous low-level messaging system" by Gavin Wood. This protocol can be necessary for the execution for certain transactions but can also be the focus for the internal workings of application. Over the years there has been a struggle in mainstream adoption of decentralized applications and one of the reasons is that applications often need to be able to exchange transient messages but it doesn't make sense to make use of a blockchain for this. That is where Whisper comes in as a decentralized messaging protocol. However, in the past the protocol hasn't received the necessary attention and at the moment it isn't really as scalable as it should be for the applications that are currently being developed. Developers are now working on the Whisper 2.0 version that should allow for this scalability and incorporate the necessary privacy precautions so that centralized infrastructure providers aren't able to monitor the information that is being exchanged between these decentralized applications. Currently, a setup is ongoing to gain support for further development between researchers, protocol implementers and application builders.[259] But how does this protocol look like? It consists of three

---

[259] https://github.com/w3f/messaging/.

elements called "envelopes," "messages," and "topics." The envelopes are packets that contain several pieces of information:

- time-to-live (in seconds);

- expiry (in Unix time);

- topics (hash tags, hashed public key of recipient with session nonce, …);

- nonce (provides proof of work requirements); and

- message data field (the encrypted payload combined with flags and signature).

Based on the nonce and the work performed on the message, there could be a prioritization of the messages that are received. On top of that, nodes can choose which privacy or performance features they prioritize and accept or reject. An example of an application running on top of Whisper is Status.im. Currently in beta, it is an app for both desktop as smartphone which integrates a messaging service and other existing DApps on one location.[260] The main goal of the newfound browser is to lower the threshold to make use of Ethereum and the decentralized applications that run on top of the network.

3.28.2   ETHEREUM SWARM

Similar to Whisper, Swarm is the solution that has been put forward by Ethereum to offer a decentralized and distributed storage platform. It is another layer in the native web3 stack that is proposed and supported by the community. The main goal of swarm is to provide a decentralized and redundant store of Ethereum's public record, in particular to store and distribute DApp code and data as well as blockchain data.[261] Some of the services that are offered or are under development are messaging, data streaming, p2p accounting, mutable resource updates, storage insurance, proof of custody, scan and repair, payment channels, and of course database services. On top of that it is DdoS resistant, censorship resistant, and promises high availability. This offers a lot of possibilities for the developers of decentralized applications while for the users there changes nothing compared to the normal use of the World Wide Web. Several APIs are offered to developers such as CLI, JSON-RPC, HTTP interface, and JavaScript. Just as with Whisper, the Swarm implementation is still under development and one should use the protocol with care. Proof of concept version 0.4 has been released in May 2019 which has provided a stable deployment infrastructure and stable testnet combined with file sharing, access control, and notifications. Future updates will bring push syncing, upload progress bars, redundancy with erasure coding, pinned content, proof of custody

[260] Jankov, T. (June 1, 2018) Ethereum messaging: explaining whisper and status.im. *Sitepoint*. https://www.site-point.com/ethereum-messaging-whisper-status/. Accessed January 13, 2020.

[261] https://swarm-guide.readthedocs.io/en/latest/introduction.html.

challenge protocol, and more. Still it can be interesting for developers to test the implementations of Ethereum Swarm in the development of DApps.

3.28.3   IPFS

"Interplanetary File System" (IPFS) is a distributed system for the storing and accessing of files, setup of websites, applications, and data in general. It is similar to Ethereum Swarm as it also wants to offer a decentralized storage layer and a content delivery protocol for decentralized applications. Just as Swarm, IPFS wants to offer an incentivization layer for the participating nodes to encourage participation and insurance to users. Similarly, the storage model used is a block model that chops up large documents into pieces that can be fetched in parallel.[262] Both IPFS and Swarm also make use of content addressing. What is content addressing? Well, standard computer users are used to location addressing where a user will type in an URL and expects a webpage based on that URL. In content-based addressing you can find web pages based on the content instead of the location. The URL of IPFS for example contains a hash of the content of the web page you are accessing. This way you can verify if what you are accessing is truly what you have asked for. Finally, there is transparent and efficient mapping of file system directories. So why create two different implementations that have the same goals? Well, there are some differences between the projects that will probably keep them both alive. First of all, the development (and the adoption) of IPFS is much further along than that of Ethereum Swarm. Swarm, on the other hand, has a stronger relation with Ethereum bringing advantages such as the live network of users, funding from the non-profit behind Ethereum and the strong ecosystem it could be implemented in. Secondly, there is a "philosophical" difference between the two projects. Swarm is part of the Ethereum and Whisper development stack for Web 3.0 and focuses on privacy, censorship resistance and is developed specifically for the needs of the Ethereum ecosystem. IPFS on the other hand is developed to be open for any protocol that wishes to develop toward Web 3.0 and therefore it also offers options such as blacklisting and source filtering. Finally, there are also some technical differences between the two projects such as different network communication and peer management protocols. Swarm makes use of the same protocols that are used by the Ethereum network. IPFS, on the other hand, makes use of libp2p network layer. Another difference is that you can upload to Swarm and use it as a cloud hosting provider while IPFS requires you to have the file on your hard drive. Closely linked to the implementation of IPFS is IPNS or "Interplanetary Name System" and is a system for creating and updating mutable links to IPFS content.[263] The name specifically is a hash of a public key and takes the form of: /ipns/QmSrPmbaUKA3ZodhzPWZnpFgcPMFWF4QsxXbkWfEptTBJd. A similar

---

[262] https://github.com/ethersphere/swarm/wiki/IPFS-&-SWARM.
[263] https://docs.ipfs.io/guides/concepts/ipns/.

implementation that currently still works faster (and is more human readable) is DNSlink. It uses the domain name instead of the hashed public key and comes closes to what users are used to today.

Filecoin

Filecoin is a project that is linked to IPFS and is the incentivization layer that enables the participants in the protocol. This implementation will make sure that there is an ongoing compensation for those participants that provide storage for the standard users of IPFS. Proof of retrievability mining is the consensus protocol in use here and is a positive reinforcement protocol while Swarm on the other hand has also some punitive measures in place to makes sure that the participants remain truthful to the purpose of the protocol.

The cryptocurrency behind Filecoin is also freely interchangeable. This is used as an extra incentive to convince people to open up storage that is currently not being used. Some part of it is aimed at creating a decentralized market for storage. If the reader remembers correctly (and has read the section on Bitcoin) some of the goals behind Filecoin are similar to Siacoin. The difference between the two is that Filecoin is really focused on IPFS as well, while Siacoin is a separate implementation in the world of Bitcoin. The Filecoin network makes use of two proof-of-storage protocols: proof of replication which allows participants to prove they have replicated the data on their own storage device and proof of space-time which is a proof of work consensus protocol that allows participants to proof that they have stored some data over a certain amount of time.[264] Their implementation also makes use of zk-SNARKs to improve the security over the network. The idea is to create a market where participants can offer space or request it and when there is a match on pricing, the order can be carried out. Similarly, when the user wants to request a file it has put on decentralized storage, there is an order matching and settlement phase. This is why there are retrieval miners and storage miners with each their own responsibilities toward the participants that make use of the storage facilities. The Filecoin network also wants to provide the capabilities for a smart contracting platform, even though it is rather limited in its setup. The platform supports basic "put" and "get" requests but file contracts could be developed in the future that would allow participants to code under which circumstances they are willing to provide storage. There is also the idea to implement smart contracts that are more generic and allow for asset tracking, naming systems and more. However, still a lot of work must be done to make Filecoin completely ready for wide use such as: specification of the Filecoin state tree in every block, full implementable Filecoin protocol specification, SNARK/STARK implementation, and more.

---

[264] https://filecoin.io/filecoin.pdf.

3.28.4 LIBP2P

Libp2p is a networking stack and library that has been modularized out of the IPFS project and is now open to use for other tools and implementations. The networking protocols used by Ethereum are largely based on the research and work done on libp2p and it is an implementation that means to facilitate the use of peer to peer networks. It has implementations that are being supported in Go, JavaScript, Rust, and Python and covers a wide arrange of services such as discovery and identification of peers, plaintext protocols that are insecure but can be interesting for certain networks, but also implementations for communication based on pre-shared keys, circuit switching protocol, TLS handshake, and transport security protocol.[265]

3.28.5 IPLD

Interplanetary Linked Data (IPLD) is closely linked to IPFS and also under development by protocol labs. It aims to bring a data model that can support the content-addressable web as it is created in the world of Web 3.0 and IPFS. The idea is that hash-linked data structures can be treated as subsets of a unified information space, unifying all data models that link data with hashes as instances of IPLD.[266] The implementation of IPLD would allow for content-addressing across blockchain networks and protocols (as long as it has some basis in hashing for content-addressing). When this would be in use, it would allow commits to a git branch to be referenced by, i.e., a Bitcoin transaction as a timestamp or allow Ethereum contracts to reference media on IPFS. In Go you can already find packages to support Git, Bitcoin, Ethereum, and IPFS but also in JavaScript a lot of implementations can be found.

3.28.6 MULTIFORMATS

The final addition to the next generation of web applications that is being offered by Protocol labs is multiformat protocols. What are multiformat protocols, you ask? Good question. The main goal of this implementation is that there is a set of protocols that aim to future-proof systems by enhancing format values with self-description. The self-description of the protocols has to adhere to a specific set of rules: they must be in-band with the values, must help to avoid lock-in and promote extensibility, must have a binary-packed representation and must have human-readable representation.[267] Currently, there are already the following implementations:

- **Multihash:** self-describing hashes;

- **Multiaddr:** self-describing network addresses;

---

[265] https://github.com/libp2p/specs.
[266] https://ipld.io/.
[267] http://multiformats.io/.

- **Multibase** : self-describing base encodings;

- **Multicodec:** self-describing serialization;

- **Multistream:** self-describing stream network protocols;

- **Multigram:** self-describing packet network protocols; and

- **Multikey:** self-describing cryptographic keys and artifacts.

Currently IPLD, IPFS, and libp2p is making use of multiformats.

3.28.7 0X PROTOCOL

The 0x protocol was developed to facilitate the exchange of Ethereum-based tokens (ERC-20 and ERC-721). The goal is to increase the liquidity in these tokens and assets so that businesses can integrate these new ways of payment into their current portals and way of working. By making a modular and user-friendly design, it can be integrated without extensive extra development. An extensive set of APIs eases developers into the use of this layer. The use cases advertised on the webpage of the 0x[268] protocol go into games and collectibles as the implementation can also be used for non-fungible tokens (remember the crypto kitties!). It facilitates the trade of these tokens in the marketplace. There are also the prediction markets. Several examples can be found in this book when it comes to decentralized prediction markets where financial stakes are represented by tokens which in turn can be traded. The 0x protocol can help making these markets more liquid as these tokens can more easily be traded with this implementation. A third example are order books that can be facilitated, just as decentralized loan markets that can gain increased liquidity with the buying and re-selling of loans in the form of tokens. Finally, there are the stable tokens which require efficient and liquid markets to succeed in their efforts. 0x can also play here a crucial role for these tokens to succeed in their efforts.

3.28.8 DAT PROTOCOL

The dat:// protocol is a peer to peer protocol that was developed especially for decentralized networks by a strong developer community.[269] It is a protocol that must help the sharing of data directly between computers. The protocol even works in networks with poor connectivity and even offline. It is able to handle large datasets and you can add or modify data while keeping a full history on your machine. it is very user friendly so that it is also accessible to people who don't necessarily have a deep technical knowledge. The protocol is in use for websites but also for art, music releases, chat programs, and more.

---

[268] https://0x.org/why#benefits.
[269] https://dat.foundation/.

How does a dat URl look like? Well, there are three main parts:

Dat://668f8d955175f92e4ced5e4f5563f55bvch0c86cc6f670352c451233777ab879/
welikedat.gif

The protocol is identified by **dat://**. Even if you are not an IT professional, you will recognize the format from http://; it is just replaced with another protocol, one suitable for the decentralized world. The second part consists out of a long list of letters and numbers, representing the ed25519 public key in a hexadecimal format. This public key allows participants to find and discover others that have the data and verify that the data wasn't changed while passing through the network. Finally, there is the suffix, which is an optional path to data within the Dat. It will most often look a bit like a file path which you can recognize from directory structures. A browser that makes use of this structure and protocol is called Beaker. Ok, so you know now how an URL looks like in the decentralized world of Dat. But how does your client discover peers from which they can download data? There is no longer a central server that we can connect to, as we are used to in the world of http. The solution comes in the form of discovery keys. Based on the public key of the Dat you are looking for, you can easily calculate the discovery key; however, the reverse isn't possible making it impossible for malicious participants to discover the URL you are looking for.[270] The discovery key is produced by making use of the BLAKE2b hashing function to hash the word "hypercore." Peers trying to reach a certain dataset or website will broadcast Dats in their local network. In the future, hyperswarm will be used to improve the process of finding and connecting with peers in the network so that data can be found even faster and shared among the participants.[271] Currently, there is a multicast DNS which are similar to regular DNS queries, except that they are broadcasted throughout the network. The packet itself is an UDP-packet that is sent to the special broadcast MAC 01:00:5e:00:00:fb and IP address 224.0.0.251 with source and destination port 5353. The DNS header looks like this for a client asking for peers:

Transaction ID	Flags	Questions	Answers	Authority Records	Additional Records
0 0	0 0	0 1	0 0	0 0	0 0

While the response from a peer that he is interested in the same Dat looks like this:

Transaction ID	Flags	Questions	Answers	Authority Records	Additional Records
0 0	132 0	0 1	0 1	0 0	0 0

---

[270] Based on traffic, attackers might still figure out how many Dats are popular, their size, IP addresses, traffic times, and volumes.

[271] https://github.com/hyperswarm.

Responses can eventually contain two TXT records: the token record and the peers record. The first is a random value that makes sure that clients don't connect to themselves. The second is a base64-encoded list of IP addresses and ports of peers interested in the same Dat. However, there is also still a centralized approach to DNS discovery which introduces once again a single point of failure but helps for a fast and global reach. To date, this is the server discovery1.datprotocol.com with a fallback discovery2.datprotocol.com. Peers have to re-announce themselves every 60 sec to remain connected and the server also cycles its tokens so that peers have to remember their last received token and update it when necessary. The destination IP address will change to 45.55.78.106 (the first discovery server) and the destination port will be 53. The peer announce request to the server has the following header:

Transaction ID	Flags	Questions	Answers	Authority Records	Additional Records
189 150	1 0	0 1	0 1	0 0	0 1

While the response from the discovery server looks like this:

Transaction ID	Flags	Questions	Answers	Authority Records	Additional Records
189 150	128 0	0 1	0 1	0 0	0 0

Finally, the SRV push notifications look like this:

Transaction ID	Flags	Questions	Answers	Authority Records	Additional Records
135 100	1 0	0 0	0 0	0 0	0 1

When our client has discovered a peer and the port number, it will open a TCP connection. The first message that is send after opening the connection is a feed message. This message consists out of a discovery key and a nonce that is randomly generated for the TCP connection. The second message that is generated will always be a handshake message that is sent on each side of the connection. It looks like this:

Number	Name	Description	Type
1	ID	Random ID so that peer doesn't connect to itself	Length-prefixed
2	Live	0 (end connection) or 1 (keep open)	Varint
3	User data	Arbitrary bytes for higher level applications	Length-prefixed
4	Extensions	Name of the extensions the peer wants to use	Length-prefixed
5	Acknowledge	0 or 1 (no need to acknowledge / acknowledge)	Varint

To prevent eaves dropping, the connection is encrypted after the second message with the XSalsa20 encryption cypher. The messages that are being relayed between the peers have the following structure until the end of the connection: length, channel, and type, followed by the actual body of the message (this is the structure of the so-called wire protocol). This is repeated over and over as long as the connection lasts. The length is the number of bytes until the next length field (and thus the next message). The channel and type consists out of 11 bits that encodes the channel number (up to a maximum of 127) and the message type (up to a maximum of 15). With the same TCP connection you can communicate with several Dats, starting the number of channels at 0 for the first Dat and so on. For the types, you can refer to the following table:

Type	Name	Description
0	Feed	I want this Dat
1	Handshake	Negotiate on TCP connection
2	Info	Start/stop downloading/uploading
3	Have	I have the data you want
4	Unhave	I no longer have the data/I didn't save the received data
5	Want	I want this data
6	Unwant	I no longer want this data
7	Request	Please send me this data
8	Cancel	Don't send me
9	Data	Here is the data
10 – 14	N/A	Unused
15	Extension	Message is not part of core protocol

Finally, there is the body of the message which is the actual content of the message. The first two fields in the message are varints or variable length integers. They are easy in use as they can encode with few bytes small numbers and can be expanded based on the need. Encoding and decoding, on the other hand, takes a bit more time compared to normal integers. The body consists out of field tags that are each time followed by values. The field tag consists out of two important parts: the field number (determined by the most significant bit) and the field type. Because the field number can have a variable length, the field flag is also a varint. In the network, the peers also send keepalive messages. These are completely empty and are discarded upon arrival but are purely used to keep the network alive in case there are TCP connections being cut when they haven't send data for a while. But what if you want a certain dataset from another peer? Say welcome to the "request"-message!

Number	Name	Description	Type
1	Index	Number of the chunk to send	Varint
2	Bytes	Ignore index if this is present, looks for specific byte	Varint
3	Hash	0/1 (send data and hashes/send only hashes)	Varint
4	Nodes	0/1 (send all hashes to verify the chunk/send only the data)	Varint

And canceling the request has the following format:

Number	Name	Description	Type
1	Index	Number of the chunk to cancel	Varint
2	Bytes	Ignore index field and cancel specific byte	Varint
3	Hash	Cancel the hash you are forwarding	Varint

The data that is being exchanged between the peers consists out of variable-length data chunks. When there is an existing Dat, new chunks can be added at the end but existing chunks can't be deleted or modified. To make sure there haven't been any modifications, there are also hashes that help verify the data hasn't been tampered with. There is also a tree like structure (Merkle tree) so that the sets of data chunks can also be verified. The tree consists out of the root hash, the parent hashes (two or more linked), and eventually the chunk hashes that verify a single chunk of the dataset. The dataset is being sent to other peers based on the "want/unwant" and "have/unhave" messages. The data message looks a bit like this:

Number	Name	Description	Type
1	index	Chunk number	Varint
2	Value	Contents of the chunk	Length-prefixed
3	Nodes	This is repeated for each hash that verifies the chunks integrity	Length-prefixed
4	Signature	Ed25519 signature of the root hash for this chunk	Length-prefixed

Dats make use of two coupled fields to represent both files and folders: metadata feed and the content feed. The metadata feed contains names, sizes, and other metadata for each of the files while the content feed contains the actual file content.

Several future updates and implementations are currently being worked on to improve the Dat protocol. The hyperdrive (the folder and file system) will completely change as they want to start working with a prefix tree to make it faster. There is also hyperswarm (as mentioned before) which will introduce new discovery mechanisms in the network. Next, there is multi-writer which will allow Dats to be updated by several authors and devices at the same time. Finally, there is the NOISE protocol which has to fix the current situation where connections can be eavesdropped.

3.28.9 CRYPTOCURRENCY IMPLEMENTATIONS ON ETHEREUM

So, we have learned thus far that Ethereum opened the door to a completely new world of smart contracts and decentralized applications. However, Ehtereum also opened the door to something else that we all know as cryptocurrencies that run on top of the network. Some tokens have their own blockchain but a lot of them run on top of Ethereum. For this, several standards were developed over the years and even tokens that are independent from the network tend to follow these standards as they are now generally accepted by the broader community.

3.28.10 EIP 20: ERC-20 TOKEN STANDARD

ERC-20 is the first of the standards that I am presenting to you as it was the first standard that was released to define the design of a token in a smart contract on the Ethereum network and was originally proposed by Fabian Vogelsteller on November 19, 2015. At the time of writing, there are almost 200,000 tokens on the Ethereum main network that comply with the ERC-20 standard. Paradoxically enough, Ether itself does not conform to the ERC-20 format so that it has the "wrapped" and converted to WETH which can then be held in a smart contract and a 1:1 peg to ether. Steps are being undertaken to change this so that ether can directly be used for the ERC-20 tokens. Until that time, Radar relay,[272] and 0x protocol[273] offer interfaces that allow you to trade directly between WETH and ERC-20 tokens. The following methods are defined in the standard and can be found on the Ethereum EIP website:[274] An important first notice is that callers also must be able to handle "false" responses from Boolean outcomes.

- Function name() public view returns (string);

- Function symbol() public view returns (string); and

- Function decimals() public view returns (uint8).

These first three functions are optional and meant to improve the usability of the token. The standards website states that interfaces should not expect these values to be present but nowadays it is almost a given that for a token, one can call the name and the symbol. It also doesn't seem a big stretch that you would like to see how many decimals a token is using. This determines how far this token can be split and used in real-life transactions.

---

[272] https://radarrelay.com/.
[273] https://0x.org/portal/account.
[274] https://eips.ethereum.org/EIPS/eip-20.

- Function totalSupply() public view returns (uint256); and

- Function balanceOf(address _owner) public view returns (uint256 balance).

These two functions are mandatory and call for the total supply of the token and the balance of a specific address.

- Function transfer(address _to, uint256 _value) public returns (bool success);

- Event transfer(address indexed _from, address indexed _to, uint256 _value); and

- Function transferFrom(address _from, address _to, uint256 _value) public returns (bool success).

The transfer function transfers a certain amount "_value" to the address "_to" and calls the Transfer event. Similarly, the transferFrom function transfers an amount "_value" from "_from" to a certain address "_to." The transferFrom function is a function used in case of a withdraw workflow, i.e., with a contract to transfer tokens, and should throuw unless the "_from" is specified and has authorized the sender of the message. Important note is that transfers with a value of 0 should also trigger the transfer event.

- Function approve(address _spender, uint256 _value) public returns (bool success); and

- Event approval(address indexed _owner, address indexed _spender, uint256 _value).

The approve function allows the "_spender" to withdraw from the account up to "_value."

- Function allowance(address _owner, address _spender) public view returns (uint256 remaining).

Finally, the allowance function defines the "_spender" that is allowed to withdraw from "_owner." Critical flaw in all of this: the difference between transfers to someone's address and to a smart contract. If you want to deposit in a smart contract, you must make use of the "approve" and "transferFrom" functions while the "transfer" function is used for a deposit in a standard wallet. If you use the "transfer" function to transfer to a smart contract, the transaction will succeed but the recipient will not be able to receive the tokens, and therefore they will be lost in limbo. Several implementations have already been created such as the OpenZeppelin implementation and the ConsenSys implementation.

3.28.11 ERC-223

The ERC-20 standard is easy to understand and therefore often used by participants to create tokens but it also has a set of flaws. One of these flaws is that once tokens are lost because they are send to a smart contract by the process that should send them to a wallet (ordinary address account). This has already resulted in millions of dollars in losses. ERC-223 was proposed by u/Dexaran and is specifically aimed at improving the ERC-20 standard and has taken care of this nasty problem by throwing an error in case of invalid transfers and canceling the transaction so that no tokens are lost. There is also the addition of an extra data parameter to the transfer function to allow for more than only token transfers. On top of that, it introduces process efficiency and reduces the gas needed to make transfers.[275] And even with these advantages, it remains backward compatible with ERC-20 not secluding the tokens based on the ERC-20 standard. The main assumption in this standard is that there is a "tokenFallback" function in the smart contract on which his "transfer" function is based. This means that the "transfer" function checks whether the receiving address is a smart contract and if this is the case, assumes that the "tokenFallback" function is there. However, if this is not the case, the tokens can still be lost!

3.28.12 ERC-721 [276]

ERC-721 is a free open standard that describes how one can create non-fungible tokens on the Ethereum blockchain. It is the format of a smart contract that allows you to securely manage, own, and trade these tokens while at the same time leaving space for extra metadata or supplemental functions.

The most famous example of ERC-721 are the cryptokitties from Axiom Zen which was released end 2017 and some of these kitties were sold for over $100,000! Today it is used by many platforms to create unique tokens representing real assets. The template can simply be copied and adjusted from the GitHub page so that it can be used in the correct fashion.[277] Below we will show a short example on ERC-721.

3.28.13 ERC-777

Another standard that is aiming to improve the ERC-20 token standard is the ERC-777 standard. This standard makes use of an alternative way to recognize the contract interface: the central registry of contracts on the Ethereum network that was introduced and defined in ERC-820 (contract pseudo-introspection registry). Every participant can make use of this interface to see if a contract makes use of certain functions or not. You could in theory create a token based on ERC-20 integrated with the functions provided by the ERC-777 standard. This would lead to positive network effects and of course to faster adoption by the participants and the community as well. Exchanges i.e., can easier support this standard than ERC-223 because of these integration possibilities.

---

[275] Wiigo Coin (January 2, 2019). ERC223 token standard pros and cons. *Medium*. https://medium.com/@wiigocoin/erc223-token-standard-pros-cons-93a01f0239f. Accessed January 14, 2020.

[276] erc721.org/.

[277] https://github.com/OpenZeppelin/openzeppelin-solidity/blob/master/contracts/token/ERC721/ERC721.sol.

3.28.14 ERC-827

The ERC-827 token standard is an extension of the standard interface of ERC20 tokens so that the execution of calls inside transfer and approvals are possible.[278] This means that the token proxy can execute a function in the receiver contract after the transfer is approved. To accomplish this, a proxy contract is used to forward the calls from the token contract. It adds three methods to the standard ERC20 methods that are already in use:

- approveAndCall: it only allows the receiver contract to use approved balances;

- transferAndCall: there is no check if the transferred balance is in fact the correct one; and

- transferFromAndCall: same as transferAndCall, allowing contracts to transfer tokens on your behalf before execution of a function.

3.28.15 ERC-664

The ERC-664 token standard wants to adapt the ERC-20 tokens so that the user balances are abstracted away from the business logic. An entire set of functions and methods are defined on the ERC page, which is still open at the time. [279]

3.28.16 ERC-677

With ERC-677, there is the introduction of functionality to allow the transfer of tokens to contracts and have the contract trigger logic on how to respond to receiving tokens within a single transaction.[280] It enters a new transaction type called "onTokenTransfer" and wants to solve the vulnerability that still persists with the ERC-223 standard. It is meant as a transitional measure toward the wider adoption of the ERC-223 tokens.

3.29 ETHEREUM CLASSIC

We already shortly discussed Ethereum Classic (ETC) in the section on the DAO but it seemed important to me to have a separate section discussing this particular hard fork of the Ethereum blockchain. While in many forms it is the same as Ethereum, there are some key differences to take into account when we look at this blockchain platform. To shortly recap, the Ethereum Classic hard fork came into being because of the aftermath of the DAO attack. Specifically, there was a part of the community that refused to participate in the hard fork that refunded the victims of the attack

[278] https://github.com/ethereum/eips/issues/827.
[279] https://github.com/ethereum/EIPs/issues/644.
[280] https://github.com/ethereum/EIPs/issues/677.

and stayed on the original chain. The first block of the Ethereum Classic chain was block number 1,920,000 on July 20, 2016. On July 23, 2016, Poloniex lists the ETC-token and the price goes up to 1/3th of the ETH token. The first days and even months the community of Ethereum was in disarray and a lot of discussions were held to such an extent that some even speak of a community "war."[281] It didn't take long for the supporters of Ethereum Classic to form their own community and when block 2,050,000 was mined, there was the official "declaration of independence" on the website which stated that ETC was no longer associated with the Ethereum Foundation. In the meantime, a group called the "Robin Hood Group," which was responsible for securing about 70% of the funds of the DAO after the hack, dumped large amounts of stolen ETC on the market in an attempt to destabilize the young market of Ethereum Classic. Poloniex took preventive action and froze the funds. In the months that followed, the community started to rebuild the Classic network. An important step was on August 31, 2016 when the frozen funds of the DAO attack were released to the DAO token holders and the hacker. However, the predictions seemed gloomy, the price of ETC remained quite stable. The ETC monetary policy regarding emission of tokens to align the interests of platform users, miners, investors and developers took place by the end of 2016. At block 3,000,000 there was the ETC Diehard upgrade that resolved several issues such as the possibility of replay attacks and took away the difficulty time bomb that was part of Ethereum. In 2017, the monetary policy is adapted so that there is a fixed-cap with an emission schedule similar to Bitcoin. Other innovations were the embedded SVM which allows the EVM and SputnikVM for embedded applications and support the Byzantine + Constantinople hard forks. Also the JSON RPC schema has been automated so that the operational costs related to libraries are reduced and make DApp development more efficient. Similarly, there has been research toward better tools for DApp deployment and UX research. The ETC JIT compiler has also been translated in EVM byte-code so that the execution time of programs has been reduced by three times. There has also been the Atlantis hard fork so that the Spurious Dragon and Byzantium network protocol upgrades are also available on the Ethereum Classic network. The beginning of 2020 saw the Agharta hard fork to allow the Constantinople and Petersburg network protocol upgrades on the ETC network and the next hard fork will be the Aztlan hard fork which should allow the Istanbul network upgrade on the Classic network.

---

[281] https://ethereumclassic.org/roadmap/.

CHAPTER 4

Hyperledger and DAGs

In this final chapter I would like to shortly introduce the Hyperledger foundation which has a rich set of tools and projects. At the time it is arguable the most well-known for private blockchain implementations. Next, I introduce some new and exciting implementations of DAG technology.

4.1 HYPERLEDGER

When people talk about Hyperledger they talk about a whole arrange of possible implementations as the Hyperledger project is part of the Linux foundation and hosts a lot of different blockchain solutions. We will go in-depth for all of these but I will share with you here a short overview of the (current) frameworks that are being supported by the overall project.

- **Hyperledger Besu** is an open-source Ethereum client written in Java that can run the public network but also private permissioned networks.

- **Hyperledger Burrow** is a permissionable smart contract machine.

- **Hyperledger Fabric** is a permissioned blockchain implementation with channel support.

- **Hyperledger Grid** focuses on supply chain solutions and is mainly based on web assembly.

- **Hyperledger Indy** provides a solution that can help with decentralized identity for organizations.

- **Hyperledger Iroh**a is the mobile application blockchain implementation.

- **Hyperledger Sawtooth** provides both permissioned and permissionless support for the EVM transaction family.

On top of these frameworks there is also a whole set of tools that have been created to support and aid the developers of blockchain solutions.

- **Hyperledger Aries** offers infrastructure support for peer-to-peer transactions.

- **Hyperledger Avalon** is a ledger independent implementation that extends computational trust to off-chain execution.

- **Hyperledger Cactus** is a blockchain integration tool.

- **Hyperledger Caliper** is a blockchain framework benchmark platform.

- **Hyperledger Cello** offer an as-a-service deployment.

- **Hyperledger Composer** which is a business solution to model possible blockchain networks.

- **Hyperledger Explorer** which acts as a block explorer.

- **Hyperledger Quilt** focuses on blockchain interoperability.

- **Hyperledger Transact** works on transaction execution and state management.

- **Hyperledger Ursa** which is a shared cryptographic library.

4.1.1   HYPERLEDGER BESU

Formerly known as "Pantheon," Hyperledger Besu was contributed to by PagaSys (which is the protocol engineering team at ConsenSys). Hyperledger Besu is a Java implementation of the Ethereum client.[282] It is the first implementation within the Hyperledger framework that is able to operate on a public blockchain. The platform aims to be as open as possible for both development and deployment, and therefore has a very modular built. Also, several consensus algorithms are included within the platform, such as proof of work (Ethash) and proof of authority: Istanbul Byzantine Fault Tolerance or Clique together with comprehensive permissioning schemes for private use cases. The storage of the solution is a RocksDB key-value database so that chain data can persist locally (making a division between blockchain data and the world state). As you can clearly see, everything is aimed at making the platform open for as many as possible applications. You can also clearly see this in the P2P networking (UDP-based discovery, TCP-based communication with ETH sub-protovol and IBF sub-rptocol) and the user-facing APIs (JSON-RPC over HTTP and WebSocket protocols but also a GraphQL API). Currently it is still in incubation status, but the future possibilities seem clear with this new and exciting platform.

4.1.2   HYPERLEDGER BURROW

The Hyperledger Burrow implementation has a working permissioned blockchain node that executes smart contracts following the Ethereum virtual machine specification. We can identify the following components.[283]

---

[282]   Dawson, R. and Baxter, M. (August 29, 2019). Announcing Hyperledger Besu. *Hyperledger*. https://www.hyperledger.org/blog/2019/08/29/announcing-hyperledger-besu. Accessed January 22, 2020.

[283]   https://github.com/hyperledger/burrow/tree/master.

- **Consensus engine:** that makes use of the Tendermint protocol to order and finalize the transactions.

- **Smart contract application:** based on the finalization of the consensus engine, the transactions are validated and applied to the transaction state (which consists of all accounts, both wallets and contract accounts).

- **Application Blockchain Interface (ABCI):** which is the interface used between the consensus engine and the smart contract application.

- **Permissioned Ethereum virtual machine:** based on the EVM and a permission scheme which must be matched before execution can take place.

- **Application Binary Interface (ABI):** the transactions are transmitted in a binary format to be processed by the blockchain code.

- **API Gateway:** Both JSON-RPC and REST endpoints are available to interact with the Burrow network.

4.1.3 HYPERLEDGER FABRIC

The next in line we want to discuss is Hyperledger Fabric. It provides a modular architecture which allows developers to create their own implementation and certain services such as the consensus mechanism and membership services to be plug and play.[284] The first version of Fabric was a contribution of IBM and Digital Asset.

Hyperledger Fabric is a private implementation that wants to help address such network requirements as identifiable participants, permissioned network that only has identified participants, high transaction throughput, low latency, and privacy of the actual transaction data. Opposite to most other blockchain or distributed ledger implementations, this platform does not require any native token to power contract execution. On top of the pluggable services, the smart contracts also run in separate container environments for isolation and they can be written in standard programming languages such as Golang and node.js.[285] The transactions in the Hyperledger Fabric environment follow a slightly different architecture compared to other platforms which they call "execute-order-validate." First, the transactions are executed and at that time checked for correctness. Next, it is ordered via the consensus protocol in place and finally, the transactions are validated before they are committed to the blockchain. As it is very popular to create enterprise-grade distributed ledger platforms nowadays, many tutorials can be found online if one is interested in creating and setting up a Hyperledger Fabric application. Perhaps the best place to start are the

---

[284] https://www.hyperledger.org/projects/fabric.
[285] https://hyperledger-fabric.readthedocs.io/en/release-1.4/whatis.html.

Hyperledger Fabric docs which give you a clear overview on how to set up your first network and start playing around with chaincode.[286]

4.1.4   HYPERLEDGER GRID

Hyperledger Grid is an implementation of the Hyperledger project that specifically focusses on supply chain. It aims to provide the tools and reusable code to further the development in cross-industry supply chain solutions which want to take advantage of the distributed ledger technology. Grid offers a framework to accomplish this combined with the necessary libraries and technologies. Important to know is that this implementation is currently still in its infancy as this aspect of the Hyperledger project was only accepted in December 2018. The Hyperledger Grid developers want to bring specific smart contracts and client interfaces for supply chain, combined with a modular set of domain-specific data models (based on models such as GS1), business logic, and SDKs.[287]

4.1.5   HYPERLEDGER INDY

Hyperledger Indy is specifically built to help bring decentralized identity to distributed ledgers. Similar to Hyperledger Grid, it provides tools, libraries, and reusable components for the creation and use of independent digital identities.[288] Identity on a blockchain can bring problems and concerns of its own such as privacy by design concerns and trust when it comes to sharing personal information on a ledger that cannot be altered. To address these concerns, Hyperledger Indy comes with specifications and implementations that can be applied whenever necessary and can be leveraged in implementations both inside and outside of the Hyperledger project. The Sovrin Foundation is a well-known utility for public identity that makes use of the Hyperledger Indy codebase.

4.1.6   HYPERLEDGER IROHA

Hyperledger Iroha is another blockchain platform that has been built in C++ and offers the "Yet Another Consensus" protocol combined with the BFT ordering service.[289] It offers a small set of fast commands and queries to help automate tasks and process transactions quickly when it comes to digital asset management and digital identity. Hyperledger Iroha was contributed to the Hyperledger Foundation by Soramitsu, Hitachi, NTT Data, and Colu. The Hyperledger Iroha implementation makes use of the PostgreSQL database which in itself supports complex analytics and reporting. It has more flexibility than the Hyperledger Burrow and Fabric implementations but also requires a bit more knowledge and expertise. So, if you try out the platform, be careful to check all the documentation that you can find and adapt only where you feel it is really necessary for

286   https://hyperledger-fabric.readthedocs.io/en/release-1.4/build_network.html#install-prerequisites.
287   https://grid.hyperledger.org/docs/grid/nightly/master/introduction.html.
288   https://www.hyperledger.org/projects/hyperledger-indy.
289   https://www.hyperledger.org/projects/iroha.

your own application. A couple of the applications of the Iroha implementation mentioned in the documentation are interbank settlement, central bank digital currencies, payment systems, national IDs, logistics, and similar processes.[290] An advantage over Ethereum-based systems is that there are several built in functions that allow for the creation and transfer of digital assets without the need to write own smart contracts. Another interesting aspect is that there are libraries in Python, JavaScript, Java, and Swift allowing communication with several types of applications, showing the flexibility of the framework

4.1.7   HYPERLEDGER SAWTOOTH

The last implementation (currently) is Hyperledger Sawtooth. Here the platform separates the application domain from the core system. This means that the application can be developed using any language without knowledge of the underlying system. These applications are also able to choose the transaction rules, permissioning, and consensus rules that are being used.[291] The applications that are being developed can be either business logic or a smart contract virtual machine and both of these implementations can coexist on the same ledger. Also, opposite to most blockchain systems that require serial transaction ordering, the Hyperledger Sawtooth platform makes use of a parallel scheduler. This allows for a substantial increase in the speed of the network while at the same time still preventing the double spending problem and providing multiple changes to the state of the network.

4.1.8   HYPERLEDGER ARIES

Next to the several platform options we have seen before, the Hyperleder foundation also offers an entire set of tools to assist with development. The first one of these tools that were are going to present here is called Hyperledger Aries. It is focused on creating, transmitting and storing digital credentials.[292] Aries builds on top of another Hyperledger project (Hyperledger Ursa) and was donated by the Sovrin Foundation, the Government of British Columbia, and Indy community developers. This implementation is still fully under development but will proof to be an integral part of future blockchain solutions that will focus on the identity part, mainly because this solution will be blockchain agnostic.

4.1.9   HYPERLEDGER AVALON

Another tool that is still in the incubation phase, is called Hyperledger Avalon (formerly Trusted Compute Framework or TCF) and was donated Enterprise Ethereum Alliance.[293] This tool should enable the secure movement of blockchain processing off the main chain, improving blockchain

[290]   https://iroha.readthedocs.io/en/latest/overview.html.
[291]   https://sawtooth.hyperledger.org/docs/core/releases/latest/introduction.html.
[292]   https://www.hyperledger.org/projects/aries.
[293]   https://www.hyperledger.org/projects/avalon.

throughput, transaction privacy and introduces attested oracles. With the introduction of Hyperledger Avalon, two main concerns with blockchain technology are being mitigated: scalability and confidentiality. By taking data from the blockchain network, the normal trade off would be integrity and resilience but that is why we have Avalon with the Trusted Execution Environment, multiparty compute, and zero-knowledge proofs.

4.1.10 HYPERLEDGER CACTUS

Hyperledger Cactus (formerly known as Blockchain Integration Framework) is a tool developed by Fujitsu and Accenture which allows you to integrate different blockchain networks in a secure manner. It maximizes pluggability so that ledger operations can be executed across multiple blockchain ledgers. [294]

4.1.11 HYPERLEDGER CALIPER

Hyperledger Caliper is a blockchain benchmark tool that will allow to measure the performance of a blockchain implementation based on a predefined set of use cases. It comes with a reporting engine that will show transactions per second, network latency, resource utilization, and more.[295] This project has been contributed by developers from Huawei, Hyperchain, Oracle, Bitwisr, Soramitsu, IBM, and the Budapest University of Technology and Economics. Currently, the solution can be used in Fabric, Sawtooth, Iroha, Burrow, and even Hyperledger Composer. Ethereum and other blockchain implementations will be able to make use of the benchmark tool in the near future.

4.1.12 HYPERLEDGER CELLO

Hyperledger Cello wants to make it possible to bring an on-demand, as a service, deployment model to the world of blockchain. This would greatly reduce the current efforts necessary to create and manage blockchains. The implementation aims to be as flexible as possible, focusing on baremetal, virtual machine or other containers, multi-tenant or single, and more. This project was contributed by IBM, with sponsorship from Soramitsu, Huawei, and Intel.[296] Even though this project is still in incubation, some first implementations can already be tested by developers.

4.1.13 HYPERLEDGER COMPOSER

Hyperledger composer is a tool developed for business professionals to quickly generate business networks with smart contracts and blockchain applications. It is based on the Hyperledger Fabric

[294] Klein, M. and Montomery, H. (May 13, 2020). TCS approves Hyperledger Cactus as new project. *Hyperledger.* https://www.hyperledger.org/blog/2020/05/13/tsc-approves-hyperledger-cactus-as-new-project. Accessed June 4, 2020.
[295] https://www.hyperledger.org/projects/caliper.
[296] https://www.hyperledger.org/projects/cello.

framework and can be very interesting if you want to develop proof of concepts in a quick and concise manner.[297] It allows for the modelling of your existing assets, participants, and transactions so that you can create a functional project that shows the possibilities of your use case.

4.1.14 HYPERLEDGER EXPLORER

The Hyperledger Explorer tool functions as a deployable block explorer for the projects that you are building. It allows for querying of blocks, transactions, all associated data, and network information. This implementation was contributed by IBM, Intel, and DTCC.[298]

4.1.15 HYPERLEDGER QUILT

Next in line of the tools that we are presenting, is Hyperledger Quilt. The goal of this tool is to offer interoperability between ledger systems by implementing the Interledger Protocol (or ILP) which allows for value transfers between distributed and non-distributed ledgers by implementing atomic swaps based on a single account namespace for accounts within each ledger.[299] Hyperledger Quilt was contributed by NTT Data and Ripple.

4.1.16 HYPERLEDGER TRANSACT

Hyperledger Transact wants to provide a standard interface for executing smart contracts so that that stands completely separate from the distributed ledger platform, thereby simplifying the efforts one has to do to actually create these distributed ledger platforms in the first place.[300] To implement these new smart contract languages, it makes use of something called "smart contract engines." This contribution has been made by Bitwise and Cargill. This project is still very young and has a lot of work to be done but it can already be interesting to check out the repository and test the library as far as it has been developed.

4.1.17 HYPERLEDGER URSA

Finally, these is also Hyperledger Ursa which provides a shared cryptographic library that enables people to increase the security of their projects and prevent duplicate work when it comes to cryptography implementations.[301] First of all, there is the base crypto library which contains a shared modular signature implementation that has several signing schemes and a common API. Second, there is also the Z-Mix library which offers a generic way to generate zero-knowledge proofs.

---

[297] https://hyperledger.github.io/composer/latest/introduction/introduction.html.
[298] https://github.com/hyperledger/blockchain-explorer.
[299] https://www.hyperledger.org/projects/quilt.
[300] https://www.hyperledger.org/projects/transact.
[301] https://www.hyperledger.org/projects/ursa.

4.2 DIGITAL ASSET

Digital asset is a platform that is focused on enterprise solutions.[302] At the core of the solution they offer, is DAML which is a smart-contract programming language that can be used to create digital agreements and automated transactions. DAML is an open-source programming language which aims at user friendliness so that people with little to no experience are able to create decentralized applications as fast as possible. It abstracts the underlying implementation details.[303] The language is also supported by Sextant, Hyperledger Sawtooth, Hyperledger Fabric, Corda but also Amazon Aurora, VMWare, and PostgreSQL. These multi-ledger options allow a certain level of flexibility for enterprises when they want to go into an implementation phase.

4.3 IOTA

IOTA is another form of distributed ledger technology which more than deserves our attention. The developers are aiming their efforts at the growing IoT-industry and data management for those devices. It has moved away of what we know as "blockchain" to the implementation of DAG technology: the Tangle. The Tangle makes use of the Tangle protocol and makes use of a ledger to store transactions. Each transaction can only be validated once it has been able to validate two previous transactions. As we have seen before in the explanation about DAGs, we know that these transactions are linked with each other with what is better known as "edges." IOTA also steps away from the miners that you can normally find in blockchain networks. Instead, each participant that transacts takes up the role of miners as two previous transactions have to be validated to make us of a proof of work calculation. This means that transactions can have two different states: pending or confirmed. These transactions aren't just propagated through the network but are actually bundled together. These bundles consist of data or instructions for a node to send tokens to another address. Important is that these bundles need complete consistency: either each transaction in the bundle can be validated or none of them can. There is no middle way. Each of these bundles has the same standard structure: head, body, and tail. The tail is index 0 and the head is the last transaction that we can find in the bundle. The head transaction of the bundle is connected again with the tails of two other bundles in the Tangle.

---

[302] https://digitalasset.com/.
[303] https://daml.com/features/.

Table 4.1: Transaction			
Field	**Description**	**Type**	**Length (trytes)[304]**
signatureMessageFragment	Signature/part of signature	String	2187
address	Sender/recipient address	String	81
value	IOTA tokens	Integer	27
obsoleteTag	User-defined (to be removed)	String	27
timestamp	Unix epoch	Integer	27
currentIndex	Index of transaction in bundle	Integer	9
lastIndex	Index of last transaction in bundle	Integer	9
bundle	Hash of bundle	String	81
trunkTransaction	Transaction hash of parent	String	81
branchTransaction	Transaction hash of parent	String	81
attachmentTag	User-defined	String	27
attachmentTimestamp	Unix epoch	Integer	9
attachmentTimestampLowerBound	Lower limit attachmentTimestamp	Integer	9
attachmentTimestampUpperBound	Upper limit attachmentTimestamp	Integer	9
nonce	X times transaction hash to check PoW	String	27

Bundles can remain pending for a while due to stress or an increased load on the network. Several actions can be taken to ensure that a bundle is confirmed and becomes part of the Tangle. First of all, you can promote your bundle by increasing the cumulative weight of its tail transaction. You can do this by sending a zero-value transaction that references both the tail transaction of the bundle and the last milestone in the Tangle. A second approach is rebroadcasting of the same bundle to a node. You can do this if you suspect that your bundle wasn't propagated through the network. Finally, there is reattachment of your bundle. When you want to reattach, you are actually creating a new bundle that you attach elsewhere to the Tangle by requesting new tip transactions and doing the proof of work again.[305] This leads to a new hash, trunkTransaction, branchTransaction, attachmentTimestamp, and nonce. Depending on the situation you are in, one can be more interesting than the other. In case of heavy load of a node, or one going offline, broadcasting can be the easiest solution. Otherwise, promotion can be the best option unless the bundle is older than six mile-

---

[304] The trinary numeric system is another way of encoding decimal numbers. i.e., the trite-encoded character "9" stands for"0,0,0" trits and represents the decimal number "0." As you can see, there are only 27 possible tryte-encoded characters

[305] https://docs.iota.org/docs/iota-basics/0.1/concepts/reattach-rebroadcast-promote.

stones or leads to a double spend. The final option is reattachment. How does a transaction actually look like in the IOTA network? It consists out of a "trunkTransaction" field, followed by "address," "value," "obsoleteTag," "currentIndex," and "timestamp." All these transactions and their respective fields are being pushed through a sponge function to produce a 81-tryte bundle hash.[306] So each transaction has a "currentIndex" field which defines the place of the transaction in the bundle and a "lastIndex" field which defines the end of the bundle, also called the "head" of the bundle. Each of the transactions (except for the head) are connected with each other through the "trunkTransaction" field. A maximum of 30 transactions can reside in a bundle to ensure proper functioning of the network. We should also make a clear distinction between input and output transactions. Input transactions can withdraw IOTA from specific addresses but such a transaction should always have a valid signature. Important is the size of the signature. In IOTA one can define security levels to an address (either 1, 2, or 3) which leads to a different private key and signature length.

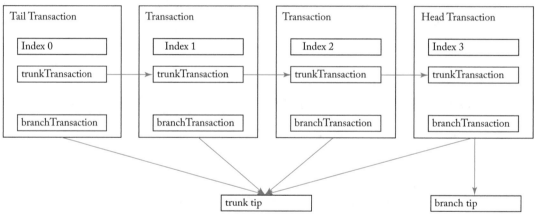

The trunkTransaction always refers to the next transaction. So, for transaction 0, this is the hash of transaction 1, and so on. The trunkTransaction of the head transaction is the same as the branchTransaction of the bundle. The branchTransaction is the tailTransaction of another bundle and is returned from the tip selection process. The branchTransaction of the head transaction is the tailTransaction from another bundle.

Figure 4.1: Structure of a bundle

Security level	Signature length
1	2,187
2	4,374
3	6,561

If this signature length is higher than security level 1, this leads to a signature that is too great to fit in a single transaction. Hence, it must be fragmented across zero-value output transactions. As an address can only be used once to withdraw from, each time all IOTA tokens need to

[306] https://docs.iota.org/docs/iota-basics/0.1/concepts/bundles-and-transactions.

be withdrawn from the address when a transaction is made. If you do not wish to send all of your IOTA tokens, you need to send the remainder to a new address. Output transactions can be divided among two main types: there are the zero-value transactions that are used to transfer parts of large signatures in the "signatureMessageFragement" field. The other type is the classic transaction which contain a certain amount of tokens that need to be transferred to a new address. Messages can be incorporated in output transactions as they don't contain a signature and therefore this space is available for messaging. Once a bundle has been created, this bundle can be validated by the node. This is done by making use of the tip selection process. This process is not enforced by the network but does rather make use of an incentive so that their own transactions have the most chance of becoming confirmed. However, these tip transactions need to be chosen at random to prevent malicious practices by the nodes involved. Before we go deeper into the process of tip selection, we first need to explain what "the coordinator" is and "milestone transactions." The coordinator is a client application that creates, signs and sends bundles of transactions from the same address at regular intervals.[307] These bundles contain milestone transactions which are used by the nodes to reach consensus. The nodes verify these transactions by making sure it doesn't lead to double spending and making sure that there is a valid signature. You can replicate the coordinator in your own private network by making use of Compass. Back to the story of tip selection, there is subgraph selection. A subgraph within the Tangle is a section of the ledger that contains all the transactions between a milestone transaction and a tip transaction.[308] The use of the subgraph is to save computational power and increase the speed of the process. What takes place next is a weighted random walk throughout the network looking for a path to a tip transaction in the network. Each of the transactions in the network receives a weight based on the "future set" which are the transactions that approve the transaction and the "Alpha configuration parameter," which is a number that affects the randomness in the network. The random walk algorithm takes these weights into account when selecting a path to new transactions. And now to the final part: addresses and signatures. Based on a seed one creates, you can start accessing addresses necessary to deal with the network. As mentioned before, each address can only be used once as an input, after which a new address needs to be generated. Based on the seed, you can generate about 9^{57} number of addresses (keep in mind you need to store the seed in a secure environment).[309] How does it do this? There is the keccak384 cryptographic hash function that takes the seed and an index and is able to produce a subseed. This subseed is then used with a cryptographic sponge function to squeeze it 27 times per security level that was chosen. This way the private key is derived from the initial seed. The public address is derived from the private key by splitting it into 81-tryte segments which are hashed 26 times. A group of 27 of these hashes is called a key fragment (based on the security level you have

[307] https://docs.iota.org/docs/the-tangle/0.1/concepts/the-coordinator.
[308] https://docs.iota.org/docs/the-tangle/0.1/concepts/tip-selection.
[309] https://docs.iota.org/docs/iota-basics/0.1/concepts/addresses-and-signatures.

1, 2, or 3 key fragments). After this process, the key fragments are hashed once to derive 1 key digest per security level and finally, the key digests are combined and hashed again to derive the final public address. While these addresses can be used as the sender address in input transactions or the receiver address in output transactions, the private keys are used for the actual signing of a transaction. When you sign a transaction, you actually sign the bundle hash of the transaction and by doing so, making it impossible for an attacker to change a transaction in a bundle without invalidating the signatures. Nodes can check the validity of a signature by using the signature and the bundle hash to find the input address of the transaction. On top of that, nodes will actually validate transactions when it received a new transaction and during the tip selection process. When we are talking about a new transaction, the node will check if the proof of work has been done, the value of the transaction doesn't exceed the total global supply, the transaction is not older than the last snapshot and not newer than two hours ahead of the node's current time and the last trit of the address is 0 for value transactions.[310] When the bundle needs to be verified, on top of the validation of the signatures, there is also a check on the value of each transaction (cannot exceed total global supply) and the total value of transactions is 0 (all spending need to go somewhere). As the network makes use of hashcash for its proof of work consensus protocol, it has a much lower difficulty when you compare the consensus protocol in place to others used by different networks such as Bitcoin.[311] The main idea is that machines are able to transact with one another by making use of the network so that cars could pay themselves for parking or houses could pay their own electricity bill. Because IOTA makes use of DAG-technology, several advantages over classic blockchain technology can be identified. First of all, there is a clear advantage when it comes to scalability. IOTA becomes faster as more participants enter the network, compared to blockchain technology that tends to slow down when the number of participants increases. Next, micro- and nano-transactions are possible in IOTA because transaction fees are no longer applicable (no more miners). IOTA is also already protected against quantum computing attacks as it makes use of Winternitz One-Time Signatures (or at least that claim is being made). Furthermore, data can be sent through communication channels between devices which are fully authenticated and tamper proof.[312] IOTA also has an own native token called IOTA tokens that can be transacted between the several nodes in the network. Below you can find the units used for the IOTA token.

---

[310] https://docs.iota.org/docs/iri/0.1/concepts/transaction-validation.
[311] https://docs.iota.org/docs/the-tangle/0.1/introduction/overview.
[312] Asolo, Bisola (November 1, 2018). IOTA explained. *Mycryptopedia*. https://www.mycryptopedia.com/iota-explained/. Accessed December 13, 2019.

Table 4.2: IOTA token units			
Name	**Unit**	**Amount of IOTA**	**Powers**
Peta IOTA	Pi	1,000,000,000,000,000	10^{15}
Terra IOTA	Ti	1,000,000,000,000	10^{12}
Giga IOTA	Gi	1,000,000,000	10^9
Mega IOTA	Mi	1,000,000	10^6
Kilo IOTA	Ki	1,000	10^3
IOTA	i	1	10^0

So far we have given an explanation of how the Tangle looks like, what transactions and bundles are and how these are signed. However, we still have something to focus on: the nodes themselves. The nodes in the IOTA network make use of what is called the IOTA Reference Implementation (IRI),[313] which is an open-source software that defines the IOTA protocol. These nodes are able to verify transactions, store valid transactions, and allow clients to interact with the IRI. For the actual creation and signing of transactions you need different client software such as Trinity (we will see an example later). Each IRI node has its own ledger with transactions that are valid and these are embedded in an append-only RocksDB database and all these databases together make up the Tangle of the IOTA network. When a node first joins the network, it starts with a process called "solidification," which means that the node starts requesting transactions that each milestone references, starting from a specific entry point milestone. For this you need to connect with other nodes that are already part of the network. You can get this from the Discord channel of IOTA where IP-addresses of nodes can be shared. Once the node has the branch and trunk of a milestone, it will start to request all the transactions that those transactions are referencing until it reaches the entry point milestone and starts again from the next one. The older the entry point milestone is, the longer the process will take.[314] Each version of the IRI software has a different starting milestone that nodes can start from and to enhance the process, the IRI node can make use of local snapshot files. The process of solidification ends when all milestones are "solid" up to the latest one. You can check this yourself if the "latestMilestoneIndex" is equal to "latestSolidSubtangleMilestoneIndex." This information can be called by making use of the "getNodeInfo" API endpoint. These IRI nodes get this information from their neighbors and also future transactions are validated by these neighbors. Consensus is achieved by all neighbors storing the same validated transactions. Also the so-called "non-solid" transactions are shared between the neighbors as they are referenced by another transaction in the IRI node's ledger. Similarly to other distributed networks, IOTA makes use of a gossip protocol to ensure communication and sharing of data. The IOTA network offers its own wallet-implementation called the "hub." It can also be integrated in application development as it

---

[313] https://docs.iota.org/docs/iri/0.1/introduction/overview.
[314] https://docs.iota.org/docs/iri/0.1/concepts/the-ledger.

covers transaction monitoring, seed creation and of course token protection.[315] Transactions made from the hub (both withdrawals and deposits) are grouped together in the same bundle called a "sweep" and the hub actively promotes and reattaches transactions until they reach confirmation. These sweeps help prevent attackers from stealing tokens from a specific spent address. The hub can support multiple users and creates seeds for each deposit address by making use of the Argon2 hash function.

4.4 HEDERA HASHGRAPH

Hedera Hashgraph is the first that is able to make use of the hashgraph technology (which is proprietary software). In theory it is able to support over 250,000 transactions a second so that scalability is no longer an issue. Every node in the network is able to "gossip" about events, which are signed pieces of information on transactions, to other nodes in the network. This gossip protocol works very efficiently and is able to spread the information throughout the network in a very fast manner. The history of the gossip protocol can be seen in a directed graph.[316]

On top of the directed graph that exists within each of the nodes, they also vote on the validity of the transactions themselves where at least 2/3 of the network is needed as witnesses. This immediately introduces the concept of "fairness." It is a non-deterministic asynchronous protocol which offers several advantages over deterministic and synchronous protocols. Deterministic protocols assume that all honest nodes will reach consensus by a certain round r for *a priori* known constant r while non-deterministic protocols do not make this assumption. On top of that, synchronous protocols assume that messages are delivered after a certain bound, while asynchronous protocols does not make this assumption. When we look at this short overview of all these features, it seems to be clear that this platform can bring several advantages when it comes to private blockchain implementations, while it would still be faced with the same challenges that "classic" blockchain implementations would have to deal with.

4.5 FANTOM

Fantom is a blockchain platform that aims to the development of smart cities but also advocates its possible use for public utilities, smart living, healthcare, education, traffic management, resource management, environmental sustainability, and more. This is another platform that brings with it the implementation of the directed acyclic graph technology (DAGs). The OPERA chain makes use of the Lachesis consensus protocol which makes use of asynchronous processing and thus leads to high processing rates without congestion. In theory, about 300,000 transactions per second are

---

[315] https://docs.iota.org/docs/hub/0.1/introduction/overview.

[316] Jia, Y. (November 8, 2017). Demystifying Hashgraph: benefits and challenges. *Hackernoon*. https://hackernoon.com/demystifying-hashgraph-benefits-and-challenges-d605e5c0cee5. Accessed October 23, 2019.

possible.[317] The Lachesis protocol consists out of events, Clothos, Atropos, and the main chain. The event block consists out of stored data which again can be made up out of multiple data packages. Next, there is the signature of the sender of the data and finally, there is the hash of the previous event block. There is also the "Clotho," which is an event block that contains a flag table. A flag table is a data structure that consists out of a Clotho index and "connectivity," which indicates the connection to other Clothos. An event block is elected as a Clotho when it can see the supra-majority (more than 2/3rd) of blocks created in the path of previous event blocks. Finally, the Clothos are used to elect the Atropos. The Atropos is a set of special event blocks and makes up the main chain. It is this main chain, the chain of Atropos blocks that is used to validate new event blocks. The network makes use of a register based virtual machine. The advantage of register-based systems over stack-based systems (which are to this day used by a majority of the platforms) is that there are no PUSH or POP instructions, which leads to less overhead when processing. Also, the reuse of stored variables is a possibility. The disadvantage is that each address must be explicitly stated.

4.6 OTHER PUBLIC AND PRIVATE PLATFORMS

The author has prepared a supplemental file to describe other public and private platforms, which the reader can download for free here: https://bit.ly/32hbjvDB or https://www.morganclaypool-publishers.com/Blockchain_Supplement.pdf.

4.7 MONERO

For the privacy lovers among you, Monero will be a well-known name. It is a cryptocurrency that has as a main focus the privacy of the participants and censorship resistance. The code base is open source and open for review, and is well maintained by 500 developers and 30 core developers. They offer quite welcoming web pages explaining the functionalities of Monero and how you can participate in the network. To achieve their goals of privacy, there was no premine or instamine, no tokens were sold nor did they have any presale. The proof of work algorithm they use is called CryptoNight which was developed in 2013 as part of the CryptoNote suite. The algorithm is based on AES encryption combined with 5 (!) hashing functions: Keccak, BLAKE, Groestl, JH, and Skein. As several of the other cryptocurrencies you will see in this book, the developers also wanted to create an ASIC-resistant algorithm, meaning that it would be difficult for specialized hardware to mine the coin, leaving some power to the CPU miners. The goal of the algorithm is to find a small-enough hash, meaning that they have to find a hash that fits under a specified threshold (sound familiar?). The input for the hash function is the block header, Merkle tree root and the number of transactions in the block. With a block time of 2 min, it well surpasses the Bitcoin network. How does the CryptoNight hash function actually work? It is a memory-hard

[317] https://fantom.foundation/contents/data/2018files/10/wp_fantom_v1.6.pdf.

hash function where the function makes use of a scratchpad with pseudo-random data It takes the input and hashes it with the Keccak hash function where the parameters are $b = 1600$ and $c = 512$.[318] From the resulting hash, the bytes 0..31 are used as an AES-256 key and expanded to 10 round keys. The bytes 64…191 from the Keccak hash are split in 8 blocks where each block is AES encrypted with the round keys (there are some differences with standard AES encryption). The resulting blocks are written in the first 128 bytes of a scratchpad, the blocks are encrypted following the same procedure again and written in the second 128 bytes. This process is repeated until 2MiB is filled. Next to this loop, there is the memory-hard loop where bytes 0…31 and 32…63 of the Keccak hash are XORed and the resulting 32 bytes are eventually used to initialize variables a and b. these variables are then used in the main loop for 524,288 times. In case the 16 byte value needs to be converted to an address in the scratchpad, it is interpreted as a little-indian integer and the 21 low-order bits are used as a byte index. However, the 4 low-order bits are used as the byte index. The read/write actions to the scratchpad happen in 16-byte blocks where each iteration is expressed by a pseudo-code. Finally, after the memory-hard loop, the bytes 32…63 from the original Keccak hash are expanded in 10 AES round keys. The bytes 64…191 from the Keccak hash are XORed within the first 128 bytes of the scratchpad. The result is encrypted just as in the first part, but with new keys, and the result is XORed with the second 128 bytes from the scratchpad, and so on. After XORing with the last 128 bytes of the scratchpad, the result is encrypted one final time, and then the bytes 64…191 in the Keccak state are replaced with this final result. The final encoded result is then passed through the Keccak-f (the Keccak permutation) with $b = 1600$. Based on the two low-order bits of the first byte of the hash are used to select a hash function: 0=BLAKE-256 [BLAKE], 1=Groestl-256 [GROESTL], 2=JH-256 [JH], and 3=Skein-256 [SKEIN]. The chosen hash function is then applied to the hash, and the resulting final hash is the output of CryptoNight. Over time the CryptoNight hash function has been modified to the current version which is also called "CryptoNightR" which is the 4th version of the hashing function. There is some criticism on the use of this hash function, as it is an expensive hash to verify, which leads to a specific vulnerability: mining nodes can be the victim of DOS-ing attacks where incorrect proofs are used on the nodes for verification. Another important point of criticism is the fact that the hash function was ultimately not able to prevent ASIC-mining. This is why the proof of work algorithm might be changed in the future but to this day, this is the hashing algorithm in place. The difficulty target changes every block, where the target is based on the last 720 blocks where the 20% of timestamp outliers are excluded. The reward for the mining decreases (just as in the Bitcoin network) but there is also a penalty in place. If you mine a block greater than the median size of the last 100 blocks, you are penalized with a lower reward. This immediately means that the block size is dynamic. The eventual supply is also uncapped, opposed to Bitcoin which is limited. How is the privacy of

---

[318] Seigen, Jameson, M., Nieminen, T., Neocortex, and Juarez A.M. (March 2013), Cryptonight Hash Function. https://cryptonote.org/cns/cns008.txt. Accessed November 12, 2019.

the users ensured? Well, the network makes use of something called ring signatures. This is used to protect the input side of a transaction, where a group of possible signers are merged together so that they can eventually create a signature to authorize a transaction.[319] This group of signers consist of the actual signer of the transaction, and a couple of non-signers which create a ring. All the participants are considered equal and valid. The non-signers are past transaction outputs that are taken from the Monero blockchain, while the actual signer uses a one-time key that corresponds with an output sent from the spender's wallet. All the inputs appear equally likely to be the output being spent to an outsider. The actual verification of transactions, to prevent double spending, is done by making use of key images. Such a key image is a secure key that is derived from the actual output being spent and is part of every ring signature transaction. It is not possible to determine which output of a ring signature actually created the key image and a list of all the used key images are being stored on the blockchain of the network. Another feature that helps to ensure privacy in the Monero world is the use of stealth addresses.[320] These are one-time used stealth addresses which further obfuscate the destination of a transaction. Finally, there is the implementation of the Kovri invisible internet project. The Kovri project is an open source network layer which allows for censorship-resistant internet use by routing traffic through nodes. The network traffic is encrypted so that your IP address cannot be linked to specific transactions.

[319] Asolo, B. (November 1, 2018). Monero Ring signature explained. *Mycryptopedia*. https://www.mycryptopedia. com/monero-ring-signature-explained/. Accessed November 26, 2019.

[320] (April 15, 2019) Liquid. https://blog.liquid.com/examples-of-privacy-coins-monero-zcash-dash. Accessed November 26, 2019.

CHAPTER 5

Some Final Remarks

I hope you found this book to be an interesting resource in the world of blockchain technology. I tried to provide a first guide which also means that I wasn't able to explain all concepts, frameworks, and networks in detail. Each of these topics deserve (and often have) complete books of their own. My goal was to open up these concepts for you and based on your interest, you can explore these further. The internet is filled with free resources that allow you to learn more but there are also numerous courses and books that give you more insight in these topics. A second remark that I would like to make concerns the fact that this is almost a "historic" artifact. This means that the second I had finished this book, there were already certain points that were outdated. An example is the Ethereum network future development plan that is open to change and has already changed. This means that historic information is exactly that. Along the same line, I apologize for the errata in this book. I wrote this after hours with a lot of enthusiasm, but I am sure that there were certain mistakes that I missed and were left behind. Based on the feedback I receive, I will adapt this book to reflect the correct information as best as possible. Finally, I would like to add that I am honored you chose to read my book and, when you read this page, to the very end. This was the first time I wrote a book like this and for me it was also a road of discovery. My writing appears sometimes to be erratic (which was sometimes a shock to me when I read back certain pages), and this is because I tried to provide as much information I possibly could while keeping the number of pages to an acceptable amount. Forgive me if certain aspects seem to be unclear, and do not hesitate to reach out. Again, based on your feedback I would like to rewrite and release a second edition. Thank you for being my reader and hopefully I will hear from you soon!

Bibliography

Antonopoulos, A. (2017). *Mastering Bitcoin: Pprogramming the Open Blockchain.* 2nd ed. California: O'Reilly Media.

Antonopoulos, A. and Wood, G. (2018). *Mastering Ethereum.* 1st ed. O'Reilly Media.

Asolo, B. (October 30, 2018). X11 algorithm rxplained. *Mycryptopedia.* https://www.mycryptopedia.com/x11-algorithm-explained/. Accessed November 27, 2019.

Asolo, B. (November 1, 2018). Monero ring signature explained. *Mycryptopedia.* https://www.mycryptopedia.com/monero-ring-signature-explained/. Accessed November 26, 2019.

Asolo, B. (November 1, 2018). What is segregated witness? *Myencryptopedia.* https://www.mycryptopedia.com/what-is-segregated-witness/. Accessed December 24, 2019.

Asolo, B. (February 16, 2019). Bitcoin Schnorr signatures explained. *Mycryptopedia.* https://www.mycryptopedia.com/bitcoin-schnorr-signatures-explained/. Accessed November 17, 2019.

Asolo, Bisola (November 1, 2018). IOTA explained. *Mycryptopedia.* https://www.mycryptopedia.com/iota-explained/. Accessed December 13, 2019.

Asolo, B. (December 20, 2018). Bitcoin's UTXO set explained. *Mycryptopedia.* https://www.mycryptopedia.com/bitcoin-utxo-unspent-transaction-output-set-explained/. Accessed November 28, 2019.

Back, A. (August 1, 2002). Hashcash–A denial of Service Counter-Measure.

Baczuk, J. (May 24, 2019) How to fork Bitcoin – Part 1. *Medium.* https://medium.com/@jordan.baczuk/how-to-fork-bitcoin-part-1-397598ef7e66. Accessed September 19, 2019.

Batiz-Benet J. (2018). go-merkledag. *Github—ipfs.* https://github.com/ipfs/go-merkledag/blob/master/README.md. Accessed August 14, 2019.

Ben-Sasson, E., Bentov, I., Hresh, Y. and Riabzev, M. (March 6, 2018). Scalable, transparent, and post-quantum secure computational integrity. *Israel Institute of Technology.* https://eprint.iacr.org/2018/046.pdf. Accessed November 23, 2019.

Bergmann, C. (April 29, 2017). The lightning network explained, part I: How to build a payment channel. *Btcmanager.* https://btcmanager.com/lightning-network-primer-pt-i-building-payment-channels/?q=/lightning-network-primer-pt-i-building-payment-channels/. Accessed July 25, 2019.

Bitcoin. (December 19, 2017). Satoshi client node discovery. *Github—Bitcoin*. https://en.bitcoin.it/wiki/Satoshi_Client_Node_Discovery. Accessed November 4, 2019.

Bitcoin. (December 13, 2019). Transaction. *Github—Bitcoin*. https://en.bitcoin.it/wiki/Transaction. Accessed December 26, 2019.

Bitcoin. (December 26, 2018). Protocol documentation. *Github—Bitcoin*. https://en.bitcoin.it/wiki/Protocol_documentation#Signatures. Accessed November 20, 2019.

Blockstream. (n.d.). How Elements works and the roles of network participants. *Blockstream*. https://elementsproject.org/how-it-works. Accessed July 13, 2019.

Blockstream Corporation. (2020). Liquid by Blockstream. *Blockstream*. Accessed July 7, 2020 https://blockstream.com/liquid/.

Bryk, A. (November 1, 2018). Blockchain attack vectors: Vulnerabilities of the most secure technology. *Apriorit*. https://www.apriorit.com/dev-blog/578-blockchain-attack-vectors. Accessed September 7, 2019.

Buterin, V. (May 9, 2016). On settlement finality. *Ethereum Blog*. https://blog.ethereum.org/2016/05/09/on-settlement-finality/. Accessed July 2, 2019.

Buterin, V. (July 11, 2014). Toward a 12-second block time. *Ethereum Blog*. https://blog.ethereum.org/2014/07/11/toward-a-12-second-block-time/#:~:text=At%2012%20seconds%20per%20block,a%20stale%20rate%20of%2050%25. Accessed July 11, 2019.

Buterin, V. (January 28, 2015). The P + epsilon attack. *Ethereum Blog*. https://blog.ethereum.org/2015/01/28/p-epsilon-attack/. Accessed September 2, 2019.

Buterin, V. (May 9, 2015). Olympic: Frontier pre-release. *Ethereum Blog*. https://blog.ethereum.org/2015/05/09/olympic-frontier-pre-release/. Accessed December 17, 2019.

Chandraker, A., Kachhela, J., and Wright, A. (2019). Digital identity, cats, and why fungibility is key to blockchain's future. *PA Consulting*. https://www.paconsulting.com/insights/blockchain-fungibility-future/. Accessed June 26, 2019.

Charlon, F. (May 13, 2015). Open Assets protocol. *Open Assets*. https://github.com/OpenAssets/open-assets-protocol/blob/master/asset-definition-protocol.mediawiki. Accessed October 17, 2019.

Chen, M. (April 13, 2019). Inter exchange client address protocol (ICAP). *Github—Ethereum*. https://github.com/ethereum/wiki/wiki/Inter-exchange-Client-Address-Protocol-(ICAP). Accessed July 3, 2019.

Chinchilla, C. (August 2, 2019). RLP. *Ethereum Wiki*. https://github.com/ethereum/wiki/wiki/RLP. Accessed December 13, 2019.

Cimpanu, C. (September 4, 2018). Bitcoin gold delisted from major cryptocurrency exchange after refusing to pay hack damages. *Zdnet.* https://www.zdnet.com/article/bitcoin-gold-delisted-from-major-cryptocurrency-exchange-after-refusing-to-pay-hack-damages/. Accessed December 19, 2019.

Conner, E. (July 27, 2018). A case for Ethereum block reward reduction to 2 ETH in Constantinople (EIP-1234). *Medium.* https://medium.com/@eric.conner/a-case-for-ethereum-block-reward-reduction-in-constantinople-eip-1234-25732431fc77. Accessed December 11, 2019.

Counterparty. (2020). The Counterparty platform. Accessed August 4, 2019 https://counterparty.io/platform/.

Couts, A. (December 27, 2013). Such generosity! After Dogewallet heist, Dogecoin community aims to reimburse victims. *Digital Trends.* https://www.digitaltrends.com/cool-tech/dogecoin-dogewallet-hack-save-dogemas/. Accessed November 29, 2019.

Curran, B. (June 26, 2018). What is Nakamoto consensus? Complete beginner's guide. *Blockonomi.* https://blockonomi.com/nakamoto-consensus/. Accessed July 12, 2019.

Curran, B. (April 18, 2020). What is practical Byzantine fault tolerance? Complete beginner's guide. *Blockonomi.* https://blockonomi.com/practical-byzantine-fault-tolerance/. Accessed July 18, 2019.

Dannen, C. (2017). *Introducing Ethereum and Solidity.* 1st ed. New York: Apress.

Dashjr, L. (January 19, 2017). Bip-0062. *Github—Bitcoin bips.* https://github.com/bitcoin/bips/blob/master/bip-0062.mediawiki. Accessed October 14, 2019.

Davies, J. (January, 2019). secp256k1. *Github—ElementsProject.* https://github.com/ElementsProject/secp256k1-zkp/tree/secp256k1-zkp/src/modules/musig?source=post_page. Accessed January 4, 2020.

Dawson, R. and Baxter, M. (August 29, 2019). Announcing Hyperledger Besu. *Hyperledger.* https://www.hyperledger.org/blog/2019/08/29/announcing-hyperledger-besu. Accessed January 22, 2020.

Decker C. and Wattenhofer R. (2015). A fast and scalable payment network with bitcoin duplex micropayment channels. *Ethz.* https://tik-old.ee.ethz.ch/file/716b955c130e6c-703fac336ea17b1670/duplex-micropayment-channels.pdf. Accessed October 13, 2019.

Decker, C. and Russell, R. (2017). eltoo: A simple Layer2 protocol for bitcoin. *Blockstream.* https://blockstream.com/eltoo.pdf. Accessed October 14, 2019.

Dexter, S. (March 11, 2018). 1% shard attack explained—Ethereum sharding (Contd.) *Mango Research*. https://www.mangoresearch.co/1-shard-attack-explained-ethereum-sharding-contd/. Accessed September 5, 2019.

Donald, J.A. (November 2, 2008). Bitcoin P2P e-cash paper. https://www.metzdowd.com/pipermail/cryptography/2008-November/014814.html. Accessed August 9, 2019.

Edmonds, R. (March 8, 2018). Best CPUs for crypto mining. *Windows Central*. https://www.windowscentral.com/best-cpus-crypto-mining. Accessed December 18, 2019.

Edwin (November 15, 2017). 1983: eCash Door David Chaum. https://www.bitcoinsaltcoins.nl/1983-ecash-david-chaum/. Accessed May 17, 2020.

Electric Coin Company. (2020). Electric Coin Company. How it works. https://z.cash/technology/. Accessed July 7, 2020.

Eyal, I. and Sirer, E.G. Majroity is not enough: Bitcoin mining is vulnerable. *Cornell*. https://www.cs.cornell.edu/~ie53/publications/btcProcFC.pdf. Accessed August 20, 2019.

Feinberg, A. (December 26, 2013). Millions of meme-based Dogecoins stolen on Christmas day. *Gizmodo*. https://gizmodo.com/millions-of-meme-based-dogecoins-stolen-on-christmas-da-1489819762. Accessed November 30, 2019.

Field, M. (November 12, 2018). Holographic consensus – part 1. *Medium*. https://medium.com/daostack/holographic-consensus-part-1-116a73ba1e1c. Accessed January 12, 2020.

Frankenfield, J. (March 5, 2018). Namecoin. *Investopedia*. https://www.investopedia.com/terms/n/namecoin.asp. Accessed November 28, 2019.

Frankenfield, J. (July 5, 2018). Peercoin. *Investopedia*. https://www.investopedia.com/terms/p/peercoin.asp. Accessed November 30, 2019.

Friedman, W. (March 26, 2015). Drop Zone: P2P E-commerce paper. https://www.metzdowd.com/pipermail/cryptography/2015-March/025212.html. Accessed August 4, 2019.

Gabizon, A. (February 28, 2017). Explaining SNARKs. *Electric Coin*. https://electriccoin.co/blog/snark-explain. Accessed December 3, 2019.

Gabizon, A. (February 28, 2017). Explaining SNARKs. *Electric Coin*. https://electriccoin.co/blog/snark-explain2. Accessed December 3, 2019.

Gabizon, A. (February 28, 2017). Explaining SNARKs. *Electric Coin*. https://electriccoin.co/blog/snark-explain3. Accessed December 3, 2019.

Gabizon, A. (February 28, 2017). Explaining SNARKs. *Electric Coin*. https://electriccoin.co/blog/snark-explain5. Accessed December 3, 2019.

Gabizon, A. (February 28, 2017). Explaining SNARKs. *Electric Coin*. https://electriccoin.co/blog/snark-explain6. Accessed December 3, 2019.

Gabizon, A. (February 28, 2017). Explaining SNARKs. *Electric Coin*. https://electriccoin.co/blog/snark-explain7. Accessed December 3, 2019.

Gabizon, A. (September 25, 2016). Zcash parameters and how they will be generated. *Electric Coin*. https://electriccoin.co/blog/generating-zcash-parameters. Accessed November 11, 2019.

Gennaro, R., Gentry, C., Parno, B. and Raykova, M. (2012). Quadratic span programs and succinct NIZKs withpout PCPs. *IBM T.J.* Watson Research Center. https://eprint.iacr.org/2012/215.pdf. Accessed November 11, 2019.

Groth, J. (October 26, 2010). Short pairing-based non-interactive zero-knowledge arguments. *University College London*. http://www0.cs.ucl.ac.uk/staff/J.Groth/ShortNIZK.pdf. Accessed October 1, 2019.

Hinkes, A. (May 29, 2016). The law of the DAO. *Coindesk*. https://www.coindesk.com/the-law-of-the-dao. Accessed December 28, 2019.

Hoogendoorn, R. (December 3, 2019). Easypaysy makes Bitcoin addresses much easier. *Medium*. https://medium.com/@nederob/easypaysy-makes-bitcoin-addresses-much-easier-faf40988614. Accessed June 4, 2020.

Hopkins et al. (1984). *The Evolution of Fault Tolerant Computing*. Springer.

Jameson, H. (November 18, 2016). Hard Fork No. 4: Spurious Dragon. *Ethereum Blog*. https://blog.ethereum.org/2016/11/18/hard-fork-no-4-spurious-dragon/. Accessed December 18, 2019.

Jankov, T. (June 1, 2018). Ethereum messaging: explaining whisper and status.im. *Sitepoint*. https://www.sitepoint.com/ethereum-messaging-whisper-status/. Accessed January 13, 2020.

Jedusor, T.E. (July 19, 2016). MimbleWimble. *Scaling Bitcoin*. https://scalingbitcoin.org/papers/mimblewimble.txt. Accessed November 26, 2019.

Jia, Y. (November 8, 2017). Demystifying Hashgraph: Benefits and challenges. *Hackernoon*. https://hackernoon.com/demystifying-hashgraph-benefits-and-challenges-d605e5c0cee5. Accessed October 23, 2019.

Jordan, R. (January 10, 2018). How to scale Ethereum: sharding explained. *Medium*. https://medium.com/prysmatic-labs/how-to-scale-ethereum-sharding-explained-ba2e283b7fce. Accessed December 21, 2019.

Kasireddy, P. (September 27, 2017). How does Ethereum work, anyway? *Medium*. https://medium.com/@preethikasireddy/how-does-ethereum-work-anyway-22d1df506369. Accessed December 13, 2019.

Klein, M. and Montomery, H. (May 13, 2020). TCS approves Hyperledger Cactus as new project. *Hyperledger*. https://www.hyperledger.org/blog/2020/05/13/tsc-approves-hyperledger-cactus-as-new-project. Accessed June 4, 2020.

Kosba, A., Miller, A., Shi, E., Wen, Z. and Papamanthou, C. (2015). Hawk: the blockchain model of cryptography and privacy-preserving smart contracts. *University of Maryland*. https://eprint.iacr.org/2015/675.pdf. Accessed November 23, 2019.

Lerner, S.D. (November, 2014). DECOR+HOP: A scalable blockchain protocol. *Semantic Scholar*. https://pdfs.semanticscholar.org/141e/d5f15e791ec7a9537a7b3250f4b7524ce302.pdf. Accessed July 27, 2019.

Liao, N. (June 9, 2017). On Settlement finality and distributed ledger technology. *Yale Journal on Regulation*. yalejreg.com/nc/on-settlement-finality-and-distributed-ledger-technology-by-nancy-liao/. Accessed June 30, 2019.

Liquid. (April 15, 2019). Liquid. https://blog.liquid.com/examples-of-privacy-coins-monero-zcash-dash. Accessed November 26, 2019.

Manning, L. (May 1, 2019). Percentage of CoinJoin bitcoin transactions triples over past year. *Bitcoin Magazine*. https://bitcoinmagazine.com/articles/percentage-coinjoin-bitcoin-transactions-triples-over-past-year. Accessed November 6, 2019.

Maxwell, G. (January 23, 2018). Taproot: Privacy preserving switchable scripting. *Linux Foundation*. https://lists.linuxfoundation.org/pipermail/bitcoin-dev/2018-January/015614.html. Accessed October 4, 2019.

Maxwell, G. (February 5, 2018). Graftroo: Private and efficient surrogate scripts under the taproot assumption. *Linux Foundation*. https://lists.linuxfoundation.org/pipermail/bitcoin-dev/2018-February/015700.html. Accessed October 24, 2019.

Mihov, D. (February 6, 2018). All Ledger wallets have a flaw that lets hackers steal your cryptocurrency. *The Next Web*. https://thenextweb.com/hardfork/2018/02/06/cryptocurrency-wallet-ledget-hardware/. Accessed September 26, 2019.

Mitra, R. (2019). Understanding Ethereum Constantinople : A hard fork. *Blockgeeks*. https://blockgeeks.com/guides/ethereum-constantinople-hard-fork/. Accessed December 20, 2019.

Monahan, T. (2017). Unprotected function. *Github—Crytic*. https://github.com/crytic/not-so-smart-contracts/tree/master/unprotected_function. Accessed September 14, 2019.

Mullins, R. (2012). What is a Turing machine? Department of Computer Science and Technolog—University of Cambridge. https://www.cl.cam.ac.uk/projects/raspberrypi/tutorials/turing-machine/one.html. Accessed June 5, 2019.

Nelaturi, K. (February 5, 2018). Understanding blockchain tech—CAP theorem. *Mangosearch.com*. https://www.mangoresearch.co/understanding-blockchain-tech-cap-theorem/. Accessed June 27, 2019.

Nopara73 (April 28, 2020). ZeroLink: The bitcoin fungibility framework. *Github—ZeroLink*. https://github.com/nopara73/ZeroLink?source=post_page. Accessed October 15, 2019.

Oscar, W. (March 22, 2019). WTF is Cuckoo Cycle PoW algorithm that attract projects like Cortex and Grin? *Hackernoon*. https://hackernoon.com/wtf-is-cuckoo-cycle-pow-algorithm-that-attract-projects-like-cortex-and-grin-ad1ff96effa9. Accessed July 25, 2019.

Payment channels. *Bitcoin*. https://en.bitcoin.it/wiki/Payment_channels. Accessed October 8, 2019.

Peterson, P. (November 23, 2016). Anatomy of a Zcash transaction. *Electric Coin*. https://electric-coin.co/blog/anatomy-of-zcash/. Accessed October 4, 2019.

Peverell, I. et al. (February 4, 2020). *Introduction to Mimblewimble and Grin*. https://github.com/mimblewimble/grin/blob/master/doc/intro.md.

Poon, J. and Dryja, T. (January 14, 2016). The Bitcoin lightening network: scalable off)chain instant payments. http://lightning.network/lightning-network-paper.pdf. Accessed October 21, 2019.

Protocol labs (July 19, 2017). Filecoin: A decentralized storage network. *Protocol Labs*. https://filecoin.io/filecoin.pdf. Accessed August 28, 2019.

Ray, J. (March 4, 2019). Sharding roadmap. *Ethereum Wiki*. https://github.com/ethereum/wiki/wiki/Sharding-roadmap. Accessed December 20, 2019.

Ray, J. (April 2, 2019). Welcome to the Ethereum Wiki! *Github—Ethereum*. https://github.com/ethereum/wiki/wiki/Ethash and https://github.com/ethereum/wiki/wiki/Dagger-Hashimoto. Accessed Augst 6, 2019.

Reiff, N. (June 25, 2019). A history of Bitcoin hard forks. *Investopedia*. https://www.investopedia.com/tech/history-bitcoin-hard-forks/.

Rosic, A. (2017). Blockchain address 101: What are addresses on blockchains? *Blockgeeks*. https://blockgeeks.com/guides/blockchain-address-101/. Accessed July 4, 2019.

Rosic, A. (2017). What is Ethereum Metropolis: the ultimate guide. *Blockgeeks*. https://blockgeeks.com/guides/ethereum-metropolis/. Accessed December 18, 2019.

Rosic, A. (2017). What is Ethereum Casper protocol? Crash course. *Blockgeeks*. https://blockgeeks.com/guides/ethereum-casper/. Accessed December 22, 2019.

Rosic, A. (2018). What is Ethereum gas? *Bockgeeks*. https://blockgeeks.com/guides/ethereum-gas/. Accessed December 11, 2019.

Roberts, D. (January 9, 2014). Mergen-Mining.mediawiki. *Github—Namecoin*. https://github.com/namecoin/wiki/blob/master/Merged-Mining.mediawiki. Accessed July 6, 2019.

Robinson, D. (2018). *ivy-Bitcoin*. https://docs.ivy-lang.org/bitcoin/language/IvySyntax.html. Accessed December 6, 2019.

Rootstock experts (2015). Sidechains, drivechains, and RSK 2-way peg design. *Rootstock*. https://www.rsk.co/noticia/sidechains-drivechains-and-rsk-2-way-peg-design/. Accessed August 12, 2019.

Schwartz, A. (January 6, 2011). Squaring the triangle: secure, decentralized, human-readable names. https://web.archive.org/web/20170424134548/http://www.aaronsw.com/weblog/squarezooko. Accessed November 28, 2019.

Schwartz, D. (August 31, 2011). How does merged mining work? *Stackexchange*. https://bitcoin.stackexchange.com/questions/273/how-does-merged-mining-work. Accessed July 10, 2019.

ScroogeMcDuckButWithBitcoin (2016). Drop Zone. https://github.com/17Q4MX2hmktmpuUKHFuoRmS5MfB5XPbhod/dropzone_ruby. Accessed August 3, 2019.

Sedgwick, K. (April 4, 2019). Decentralized networks aren't censorship-resistant as you think. *News.Bitcoin.com.* https://news.bitcoin.com/decentralized-networks-arent-as-censorship-resistant-as-you-think/. Accessed July 2, 2019.

Seigen, Jameson, M., Nieminen, T., Neocortex, and Juarez, A. M. (March 2013). Cryptonight Hash Function. https://cryptonote.org/cns/cns008.txt. Accessed November 12, 2019.

Sharma, R. (June 25, 2019). What is dash cryptocurrency? *Investopedia*. https://www.investopedia.com/tech/what-dash-cryptocurrency/. Accessed November 27, 2019.

Shead, M. (February 14, 2011). State machines–basics of computer science. *Blog.markshead.com*. https://blog.markshead.com/869/state-machines-computer-science/. Accessed June 5, 2019.

Smith, N.T. (2017). SHA 256 pseuedocode? *Stackoverflow*. https://stackoverflow.com/questions/11937192/sha-256-pseuedocode/46916317#46916317. Accessed May 26, 2020.

Sompolinsky, Y. and Zohar, A. (August, 2013). Secure high-rate transaction processing in bitcoin. *IACR*. https://eprint.iacr.org/2013/881.pdf. Accessed July 30, 2019.

Sompolinsky, Y., Lewenberg, Y., and Zohar, A. (2016). SPECTRE: Serialization of proof-of-work events: Confirming transactions via recursive elections. *HUJI*. www.cs.huji.ac.il/~yoni_sompo/pubs/17/SPECTRE.pdf. Accessed August 1, 2019.

Sompolinsky, Y., Wyborski, S., and Zohar, A. (February 2, 2020). PHANTOM and GHOST-DAG. A scalable generalization of Nakamoto consensus. *IACR*. https://eprint.iacr.org/2018/104.pdf. Accessed February 27, 2020.

Song, J. (2019). *Programming Bitcoin: Learn How to Program Bitcoin from Scratch*. 1st ed. Boston, MA: O'Reilly, p. 123.

Spilman, J. (April 20, 2019). Anti DoS for tx replacement. *Linux Foundation*. https://lists.linuxfoundation.org/pipermail/bitcoin-dev/2013-April/002433.html. Accessed October 8, 2019.

Stepanov, H. (July 1, 2019). bip-0143. *Github—Bitcoin*. https://github.com/bitcoin/bips/blob/master/bip-0143.mediawiki. Accessed December 28, 2019.

Stone, D. (March 26, 2018). An overview of SPECTRE—a blockDAG consensus protocol (part 2). *Medium*. https://medium.com/@drstone/an-overview-of-spectre-a-blockdag-consensus-protocol-part-2-36d3d2bd33fc. Accessed August 3, 2019.

Stone, D. (March 29, 2018). An overview of PHANTOM: A blockDAG consensus protocol (part 3). *Medium*. https://medium.com/@drstone/an-overview-of-phantom-a-blockdag-consensus-protocol-part-3-f28fa5d76ef7. Accessed August 4, 2019.

Sztorc, P. (December 14, 2015). Truthcoin. *Truthcoin*. http://bitcoinhivemind.com/papers/truthcoin-whitepaper.pdf. Accessed November 18, 2019.

Thake, M. (November 9, 2018). What is DAG distributed ledger technology? *Medium*. https://medium.com/nakamo-to/what-is-dag-distributed-ledger-technology-8b182a858e19. Accessed August 14, 2019.

Towns, A. (July 13, 2018). Generalised taproot. *Linux Foundation*. https://lists.linuxfoundation.org/pipermail/bitcoin-dev/2018-July/016249.html. Accessed October 10, 2019.

Towns, A. (December 14, 2018). Schnorr and taproot (etc) upgrade. *Linux Foundation*. https://lists.linuxfoundation.org/pipermail/bitcoin-dev/2018-December/016556.html?source=post_page. Accessed January 8, 2020.

Tran, A. (May 23, 2018). An introduction to the BlockDAG paradigm. *Daglabs*. https://blog.daglabs.com/an-introduction-to-the-blockdag-paradigm-50027f44facb. Accessed August 28, 2019.

Tromp, J. (November, 2019). Cuck(at)oo cycle. *Github—Cuckoo*. https://github.com/tromp/cuckoo. Accessed July 22, 2019.

Tual, S. (August 4, 2015). Ethereum Protocol Update 1. *Ethereum Blog*. https://blog.ethereum.org/2015/08/04/ethereum-protocol-update-1/. Accessed December 17, 2019.

Unibright.io (December 7, 2017) Blockchain evolution: from 1.0 to 4.0. https://medium.com/@UnibrightIO/blockchain-evolution-from-1-0-to-4-0-3fbdbccfc666. Accessed July 1, 2020

Van Hijfte, S. (2020). *Decoding Blockchain for Business*. 1st ed. New York: Apress.

Van Wirdum, A. (January 24, 2019). Taproot is coming: What it is, and who will benefit Bitcoin. *Bitcoin Magazine*. https://bitcoinmagazine.com/articles/taproot-coming-what-it-and-how-it-will-benefit-bitcoin. Accessed October 2, 2019.

Vu, Q.H., Lupu, M., and Ooi, B.C. (2010). *Peer-to-peer Computing: Principles and Applications*. 1st ed. Springer, p. 35

Wiigo Coin (January 2, 2019). ERC223 token standard pros and cons. *Medium*. https://medium.com/@wiiggocoin/erc223-token-standard-pros-cons-93a01f0239f. Accessed January 14, 2020.

Woo Kim, S. (May 28, 2018). Safety and Liveness—Blockchain in the point of view of FLP impossibility. *Medium*. https://medium.com/codechain/safety-and-liveness-blockchain-in-the-point-of-view-of-flp-impossibility-182e33927ce6. Accessed June 28, 2019.

Wood, G. (2019.) Ehtereum: a secure decentralized generalized transaction ledger Byzantium version. https://ethereum.github.io/yellowpaper/paper.pdf.

Wuille, P., Poelstra, A., and Kanjalkar.S. (2019). Analyze a miniscript. *Blockstream*. http://bitcoin.sipa.be/Miniscript/. Accessed December 7, 2019.

Wuille, P. (January 16, 2020). Bip taproot. *Github—Bitcoin bips*. https://github.com/sipa/bips/blob/bip-schnorr/bip-taproot.mediawiki. Accessed January 20, 2020.

Zander, T. (November 30, 2016) Classic is back. https://web.archive.org/web/20170202055402/https://zander.github.io/posts/Classic%20is%20Back/.

WEBSITES TO VISIT

https://forkdrop.io/how-many-bitcoin-forks-are-there.

https://github.com/bitcoinxt/bitcoinxt/releases.

https://bitcoinclassic.com/devel/Blocksize.html.

https://bitcoinclassic.com/news/closing.html.

https://www.bitcoinunlimited.info/.

https://www.bitcoincash.org/.

https://bitcoinsv.io/.

https://bitcoingold.org/.

https://www.bitcoindiamond.org/.

https://www.bitcoininterest.io/.

https://btcprivate.org/.

https://en.bitcoin.it/wiki/List_of_address_prefixes.

https://www.utf8-chartable.de/unicode-utf8-table.pl.

http://www.asciitable.com/.

https://web.getmonero.org/get-started/what-is-monero/.

https://litecoin.org/.

https://www.dash.org.

https://www.namecoin.org/.

https://knowyourmeme.com/memes/doge.

https://dogecoin.com/.

https://en.bitcoinwiki.org/wiki/GridCoin.

https://gridcoin.us/.

http://primecoin.io/.

http://blockchainlab.com/pdf/Ethereum_white_paper-a_next_generation_smart_contract_and_
decentralized_application_platform-vitalik-buterin.pdf.

https://github.com/ethereum/devp2p/blob/master/rlpx.md.

https://github.com/ethereum/wiki/wiki/Design-Rationale.

https://ethereum-homestead.readthedocs.io/en/latest/introduction/the-homestead-release.html.

http://eips.ethereum.org/EIPS/eip-608.

https://docs.ethhub.io/ethereum-roadmap/ethereum-2.0/eth-2.0-phases/.

https://education.district0x.io/general-topics/understanding-ethereum/ethereum-sharding-ex-
plained/.

https://github.com/ewasm/design.

https://education.district0x.io/general-topics/understanding-ethereum/basics-state-channels/.

http://plasma.io/.

https://education.district0x.io/general-topics/understanding-ethereum/understanding-plasma/.

https://www.sec.gov/news/press-release/2014-111.

https://github.com/w3f/messaging/.

https://swarm-guide.readthedocs.io/en/latest/introduction.html.

https://github.com/ethersphere/swarm/wiki/IPFS-&-SWARM.

https://docs.ipfs.io/guides/concepts/ipns/.

https://filecoin.io/filecoin.pdf.

https://github.com/libp2p/specs.

https://ipld.io/.

http://multiformats.io/.

https://0x.org/why#benefits.

https://dat.foundation/.

https://github.com/hyperswarm.

https://radarrelay.com/.

https://0x.org/portal/account.

https://eips.ethereum.org/EIPS/eip-20.

erc721.org/.

https://github.com/OpenZeppelin/openzeppelin-solidity/blob/master/contracts/token/ERC721/
        ERC721.sol.

https://github.com/ethereum/eips/issues/827.

https://github.com/ethereum/EIPs/issues/644.

https://github.com/ethereum/EIPs/issues/677.

https://ethereumclassic.org/roadmap/.

https://cosmos.network/intro.

https://cosmos.network/docs/intro/sdk-design.html#baseapp.

https://cosmos.network/docs/spec/ibc/.

https://tendermint.com/docs/.

https://tendermint.com/docs/spec/abci/.

https://github.com/hyperledger/burrow/tree/master.

https://www.hyperledger.org/projects/fabric.

https://hyperledger-fabric.readthedocs.io/en/release-1.4/whatis.html.

https://hyperledger-fabric.readthedocs.io/en/release-1.4/build_network.html#install-prerequisites.

https://grid.hyperledger.org/docs/grid/nightly/master/introduction.html.

https://www.hyperledger.org/projects/hyperledger-indy.

https://www.hyperledger.org/projects/iroha.

https://iroha.readthedocs.io/en/latest/overview.html.

https://sawtooth.hyperledger.org/docs/core/releases/latest/introduction.html.

https://www.hyperledger.org/projects/aries.

https://www.hyperledger.org/projects/avalon.

https://www.hyperledger.org/projects/caliper.

https://www.hyperledger.org/projects/cello.

https://hyperledger.github.io/composer/latest/introduction/introduction.html.

https://github.com/hyperledger/blockchain-explorer.

https://www.hyperledger.org/projects/quilt.

https://www.hyperledger.org/projects/transact.

https://www.hyperledger.org/projects/ursa.

https://digitalasset.com/.

https://daml.com/features/.

https://docs.iota.org/docs/iota-basics/0.1/concepts/.

https://docs.iota.org/docs/iri/0.1/introduction/overview.

https://agreements.network/files/an_whitepaper_v1.0.pdf.

https://steem.com/developers/.

https://www.steem.com/steem-whitepaper.pdf.

https://steem.com/steem-bluepaper.pdf.

https://smt.steem.com/smt-whitepaper.pdf.

https://eos.io/why-eosio/.

https://github.com/EOSIO.

https://www.goquorum.com/developers.

https://neo.org/.

https://docs.bigchaindb.com/en/latest/decentralized.html.

https://github.com/corda/corda.

https://www.corda.net/get-started/.

https://docs.corda.net/_static/corda-technical-whitepaper.pdf.

http://aeternity.com/documentation-hub/protocol/oracles/oracle_transactions/.

https://www.cortexlabs.ai/Cortex_AI_on_Blockchain_EN.pdf.

https://www.ripple.com/use-cases/.

https://www.stellar.org/how-it-works/stellar-basics/.

https://fantom.foundation/contents/data/2018files/10/wp_fantom_v1.6.pdf.

https://komodoplatform.com/antara-framework/.

https://tezos.com/.

https://www.investopedia.com/tech/what-tron-trx/.

https://lisk.io/.

https://www.celer.network/tech.html.

https://connext.network/.

https://www.counterfactual.com/technology/.

https://funfair.io/how-it-works/our-solution/.

https://raiden.network/.

https://spankchain.com/products.

https://trinity.tech/#/.

https://truebit.io/.

https://loomx.io/.

https://matic.network/.

https://alacris.io/.

https://skale.network/.

https://oceanprotocol.com/.

Author Biography

Stijn Van Hijfte has been working at Howest Applied University College since 2017, where he teaches applied computer science and is active as an expert at Deloitte. He has a background in economics, IT, and data science and is often called in as a translator between business and IT departments. He started back in 2012 with some first investigations in the blockchain space and had his entire living room looking like a science experiment to connect to the Ethereum network in 2015. His continued interest in digital solutions has led to him studying many extra certifications and destroying equipment in the process. This is his second book on blockchain technology but the first one where he wants to share technical insights and knowledge.

Printed in the United States
by Baker & Taylor Publisher Services